Bacteria-Metal Interactions

Daad Saffarini
Editor

Bacteria-Metal Interactions

Editor
Daad Saffarini
Biological Sciences
University of Wisconsin-Milwaukee
Milwaukee, WI
USA

ISBN 978-3-319-34498-0 ISBN 978-3-319-18570-5 (eBook)
DOI 10.1007/978-3-319-18570-5

Springer Cham Heidelberg New York Dordrecht London

Softcover reprint of the hardcover 1st edition 2015

Printed on acid-free paper

Springer International Publishing AG Switzerland is part of Springer Science+Business Media (www.springer.com)

Acknowledgments

The field of bacterial interactions with metals has witnessed remarkable growth, and this book is a modest attempt to present up-to-date information about bacterial metal respiration, remediation, and detoxification. This book would not have been possible without the contributions of all the authors who graciously accepted my invitation to write a chapter. Special thanks to Larry Barton who suggested that I edit this book, and for his continued support and advice during its preparation.

Daad Saffarini

Acknowledgments

Contents

Contributors

Larry L. Barton Department of Biology, University of New Mexico, Albuquerque, NM, USA

Dennis A. Bazylinski University of Nevada at Las Vegas, Las Vegas, NV, USA

Alex Beliaev Pacific Northwest National Laboratory, Richland, WA, USA

Rachida Bouhenni Division of Opthalmology Research, Summa Health System, Akron, OH, USA

Ken Brockman Nationwide Children's Hospital, The Research Institute, The Ohio State University College of Medicine, Columbus, OH, USA

Susana K. Checa Facultad de Ciencias Bioquímicas y Farmacéuticas, Departamento de Microbiología, Instituto de Biología Molecular y Celular de Rosario (IBR, CONICET-UNR), Universidad Nacional de Rosario, Consejo Nacional de Investigaciones Científicas y Técnicas, Rosario, Argentina

Brian H. Lower The Ohio State University, Columbus, OH, USA; School of Environment and Natural Resources, Columbus, OH, USA

Steven K. Lower The Ohio State University, Columbus, OH, USA

Zachery Oestreicher The Ohio State University, Columbus, OH, USA

Lucas B. Pontel Facultad de Ciencias Bioquímicas y Farmacéuticas, Departamento de Microbiología, Instituto de Biología Molecular y Celular de Rosario (IBR, CONICET-UNR), Universidad Nacional de Rosario, Consejo Nacional de Investigaciones Científicas y Técnicas, Rosario, Argentina; MRC Laboratory of Molecular Biology, Cambridge, UK

Daad Saffarini Department of Biological Sciences, University of Wisconsin-Milwaukee, Milwaukee, WI, USA

Sheetal Shirodkar Amity Institute of Biotechnology, Uttar Pradesh, India

Fernando C. Soncini Facultad de Ciencias Bioquímicas y Farmacéuticas, Departamento de Microbiología, Instituto de Biología Molecular y Celular de Rosario (IBR, CONICET-UNR), Universidad Nacional de Rosario, Consejo Nacional de Investigaciones Científicas y Técnicas, Rosario, Argentina

Francisco A. Tomei-Torres Division of Toxicology and Human Health Sciences, Agency for Toxic Substances and Disease Registry, Atlanta, GA, USA

Hufang Xu Department of Geoscience, University of Wisconsin, Madison, WI, USA

Thomas Zocco Materials Science Division, Los Alamos National Laboratory, Los Alamos, NM, USA

Chapter 1
Bacterial Copper Resistance and Virulence

Lucas B. Pontel, Susana K. Checa and Fernando C. Soncini

Abstract Copper is essential for most organisms. However, it is also toxic even at low levels, especially when its local concentration or intracellular distribution is not properly controlled. Similar to other organisms, bacteria have evolved specific copper homeostasis systems for maintaining a suitable intracellular concentration of this essential metal and at the same time, avoiding its toxic effects. Recent evidence indicates that intracellular copper actively contributes to the host innate immune response against bacterial infections and pathogens have acquired specific mechanisms to deal with this intoxicant. Here, we focus on the different arrays of metal sensing and regulatory systems employed by bacterial pathogens to mount the proper response to counteract the toxic effects of copper allowing survival and replication inside the host.

Keywords Copper homeostasis · Copper function and toxicity · Bacterial pathogenesis · Innate immune response · *Salmonella* · *Mycobacterium*

1.1 Introduction

The bioavailability of copper in earth dramatically increased after the advent of oxygen in the atmosphere. These physicochemical changes also resulted in the oxidation of sulphide group of proteins that became accessible to complex soft metals such as copper (Cu). As a consequence, many proteins evolved to use Cu as

---

L.B. Pontel · S.K. Checa · F.C. Soncini (✉)
Instituto de Biología Molecular y Celular de Rosario (IBR, CONICET-UNR), Departamento de Microbiología, Facultad de Ciencias Bioquímicas y Farmacéuticas, Universidad Nacional de Rosario, Consejo Nacional de Investigaciones Científicas y Técnicas, Ocampo y Esmeralda, 2000 Rosario, Argentina
e-mail: soncini@ibr-conicet.gov.ar

L.B. Pontel
MRC Laboratory of Molecular Biology, Cambridge CB2 0QH, UK

D. Saffarini (ed.), *Bacteria-Metal Interactions*,
DOI 10.1007/978-3-319-18570-5_1

a cofactor, integrating it into different electron transfer and metabolic pathways [1]. At the same time, the increased availability of reactive Cu species demanded the evolution of a number of strategies to handle its intracellular concentration and counteract unwanted damage [2]. This involved the development of specific sensor/ response systems to tightly control Cu uptake and removal of its excess according to the metabolic requirements. The ability to maintain the intracellular copper quota allows microorganisms to adapt to a variety of environments, and recent evidence indicates that pathogens may have evolved copper handling mechanisms to survive in the host [3, 4]. Accordingly, both the essentiality and toxicity of Cu, and the ability of the host to control Cu availability would influence host-pathogen interactions. The outcome of this balance could determine if it results either in a productive infection or elimination of the pathogen. Recent evidence also indicates that mammalian macrophages can actively accumulate Cu ions in subcellular compartments, restricting bacterial growth [3]. As a consequence, the genes involved in Cu resistance acquire particular relevance in pathogens that undergo intracellular survival and replication during their infective cycle.

In recent years, the number of newly identified or proposed Cu-containing polypeptides or proteins involved in copper handling has increased as the result of the direct detection or of newly posted genomic sequences. In this chapter, we summarize the current knowledge of the mechanisms that bacteria employ to fulfil their Cu demands and, at the same time, to defend themselves from the harmful effects of this metal, focusing on pathogens. We discuss how these pathways may serve to develop new strategies against infection diseases.

1.2 Copper as an Essential yet Highly Toxic Element

Copper is an ideal cofactor for redox enzymatic reactions because it can cycle between two oxidation states, Cu(I) and Cu(II). This distinctive attribute has made this transition metal suitable for driving many biological processes that involve single electron shuttling, such as energy transduction, iron handling, and free radical neutralization. Examples of enzymes that build their catalytic mechanism on copper are oxidases [5], in which copper catalyzes the reduction of a dioxygen molecule to H_2O_2 or to two molecules of H_2O, and oxygenases, which use copper to activate O_2 and catalyze the incorporation of one or two atoms of oxygen into organic molecules [6]. Copper is also a catalytic metal in azurins and plastocyanins, small families of proteins involved in transfer of electrons for diverse processes [7].

Despite these beneficial roles, the imbalance in copper levels can be harmful. Failure in copper homeostasiscan lead to several human diseases such as Menkes syndrome, Wilson's disease, as well as Parkinson and Alzheimer's diseases [8–10]. The toxicity of Cu has been linked to different mechanisms. Firstly, Cu(I)/(II) is at the top of the Irving–Williams series that highlights the ability of a metal ion to react with available ligands. Inside cells, Cu ions interact with sulfur, oxygen and imidazole ligands, displacing other cations from their active site in enzymes [11]. Secondly, the redox potential of the Cu(I)/Cu(II) pair is close to the redox value of the bacterial cytoplasm, which makes copper an extremely dangerous cation. Redox cycling of Cu ions can generate deleterious free radicals derived from oxygen through Fenton-like reactions, resulting in lipid peroxidation as well as protein and DNA damage [12]. Iron–sulfur clustersiron of proteins that perform key cellular metabolic functions have been also shown to be direct targets for Cu toxicity. The first observation, made in *Escherichia coli*, showed that Cu can block branched-chain amino acid biosynthesis by inactivating isopropylmalate dehydratase, an enzyme with a solvent-exposed Fe-S cluster in its active site [13]. Further studies demonstrated that Cu excess not only displaces iron from their coordinating sulfur ligands in these clusters, but also affects the formation of new iron-sulfur clusters [14].

1.3 Bacterial Control of the Copper Quota

The essentiality, and at the same time toxicity, of copper makes active handling of this metal a vital skill for most organisms. Bacteria are not different from other organisms in their requirement for copper to fulfill essential activities. However, there is little or no requirement for this metal in the bacterial cytoplasm. This is probably because most known bacterial cupropropteins are located in the plasma membrane or in the periplasm, or are secreted extracellularly [15]. Furthermore, because Cu utilization is linked to aerobic metabolism, obligate anaerobic bacteria have only few or even no cuproproteins detected to date [16].

Dedicated Cu uptake mechanisms were reported only in a few species, although the mechanism for Cu acquisition has not been completely elucidated. The membrane bound/periplasmic protein pair CopDC were described to be essential for Cu uptake in *Pseudomonas syringae* [17]. Interestingly, in *Bacillus subtilis* the homologous YcnJ protein combines both functions in a single polypeptide [18]. Other families of transporters were also linked to Cu influx. A member of the major facilitator superfamily named CcoA was found to perform this function in *Rhodobacter capsulatus* [19]. Outer membrane channels such as OmpF and the *E. coli* ComC [20], metal permeases bound to the cell membrane such as ZupT from *E. coli* [21], or even broad substrate range ABC-transporters such as MstABC from *Streptococcus pyogenes* [22], and ATPases such as HmtA from *P. aeruginosa* [23], were also found to mediate Cu acquisition. Some species were reported to secrete chelators such as copper-specific methanobactin and coproporphyrin or Fe-uptake siderophores that are essential to acquire the metal from the medium under Cu-limiting conditions [24]. In some cases they can also sequester the metal outside the cell to avoid toxicity.

No specific Cu importers or Cu scavenging proteins have been yet identified in pathogenic bacteria, suggesting that either the intracellular demand for Cu is low or the metal is readily available in the infected hosts. On the contrary, these pathogens dedicate most of their efforts to restrict the availability of Cu, using efflux , redox conversion or sequestration, as well as directing its intracellular trafficking just to the target proteins.

1.4 Chaperones, Oxidases and Efflux Systems to Control Copper Excess

Different families of proteins maintain the copper quota in each compartment. Central components of this process are Cu-chaperones, a class of small proteins that safely transport the metal from and to specific partners. The Atx1/CopZ family, probably the best characterized in this group, is responsible for cytoplasmic Cu(I) trafficking. These proteins have a characteristic MxCxxC Cu(I) binding motif and a classic βαββαβ ferredoxin-like folding [25]. In most genomes, the Cu chaperone coding gene is usually found next to a gene that encodes a P_{1B}-type ATPases, which drives copper efflux to the periplasmic space. Directional delivery of Cu(I) from the chaperone to the membrane transporter was demonstrated, indicating that these products are actually not only genetically but also functionally linked [26, 27]. Besides, the transfer of Cu(I) from the Atx1-like chaperone to the transporter located in cyanobacterial thylakoids apparently is essential for proper metallation of cuproproteins such as plastocyanin and a caa_3-type cytochrome oxidase [28]. The *Enterococcus hirae* CopZ chaperone provides Cu(I) to metalloregulatory proteins reinforcing its role in intracellular Cu handling and resistance [28]. A member of a novel class of cell membrane-anchored Cu(I) chaperones not structurally related to CopZ was recently identified in *Streptococcus pneumonia* [29]. This protein, named CupA, was proposed to increase Cu-resistance by sequestering Cu(I) in the cytoplasm and/or by delivering the metal ion to the P_{1B}-type ATPase CopA.

Chaperone-mediated copper handling is achieved not only in the cytoplasm but also in the periplasm of Gram-negative species. *E. coli* CusF is an unusual Cu-chaperone that uses one histidine, two methionines, and the aromatic ring of a tryptophan to coordinate the metal [30, 31]. This metal-binding architecture probably facilitates copper binding in the oxidizing environment of the periplasm. CusF delivers Cu(I) to CusB, the periplasmic component of the CusABC complex that mediates Cu efflux to the medium (see below) [32].

Different types of Cu chaperones are required during the assembly of cuproproteins. For example, the α-proteobacteria *Rhodobacter sphaeroides* uses three periplasmic Cu-chaperones, PCu(A)C, PrrC (Sco), and Cox11, to metallate the Cu centers of the two cytochrome c oxidases required for aerobic growth [33]. *B. subtilis* has a membrane-associated protein homologue of the yeast Sco1 that provides Cu(I) to the Cu_A site of the *caa₃* oxidase [34]. There are probably other yet unidentified Cu-chaperones required for the assembly of cuproenzymes. Recently, CueP from *Salmonella enterica* serovar Typhimurium was shown to be required for the metallation of SodCII [35]. The crystal structure of CueP has been determined revealing a V-shaped dimeric structure [36] with no homology to other known Cu chaperones.

Efflux systems were described to play a crucial part in the intracellular handling of copper. The P_{IB}-type of ATPases transport Cu(I) across cell membranes in a wide range of organisms, including bacteria, archaea, and eukarya [37]. They share the common intramembranous CPx motif (Cysteine-proline-x, where x can be cysteine, serine or histidine). In addition, they have 8 to 12 transmembrane segments with the N- and C-termini exposed to the cytoplasm, and a large cytoplasmic domain used to couple the energy provided by ATP hydrolysis to the transport of substrates. A specific feature of P_{1B}-ATPases is the presence of one to six cytoplasmic metal binding domains (MBD). The MBD harbors the metal binding signature CxxC also present in soluble Cu-chaperones [38]. Functionally, there are two groups of P_{1B}-ATPases [39]. Proteins represented by CopA from *E. coli*, direct Cu (I) transport at high rate to eliminate the toxic metal ion from the cytoplasm. The second, FixI/CopA2 group, drives Cu(I) transport at low rate and has been linked to cuproproteins metallation. As mentioned above, although there are reports on the interaction of these Cu(I) transporters with CopZ [26, 27], the strict requirement of this interaction for efficient metal efflux has not been conclusively demonstrated.

Resistance-nodulation-cell division RND protein family members are efflux systems also involved in Cu resistance in several bacterial species. The best characterized is the *E. coli* CusCFBA system [40]. This multicomponent efflux pump transports cytosolic or periplasmic Cu(I) across the inner and outer membranes of Gram negative species [41]. CusA, the RND-like protein, is the catalytic subunit, while CusC forms the outer membrane pore. The membrane fusion protein CusB not only spans the periplasmic space linking the other two components but also interacts with CusF, the Cu-chaperone, to drive Cu(I) efflux from the periplasm [42]. Contrary to P_{1B}-ATPases, these efflux systems are not widespread and some pathogenic Gram negative bacteria, such as *Salmonella*, do not harbor a Cus-like system. Interestingly, a *Salmonella*-specific gene product, CueP, can phenotypically substitute Cus, although the molecular mechanism to perform this action is still not understood [43].

The multicopper oxidases also participate in copper handling, although, they have been associated to redox modification rather than transference or pumping of Cu. The best characterized is probably CueO from *E. coli* [44]. This periplasmic enzyme harbors four copper atoms that lie in electron transfer and dioxygen reduction sites [45]. Its cuprous oxidase activity, i.e. the conversion of Cu(I) to the less toxic Cu(II), is well documented [46]. The enzyme also exerts oxidase activity against other substrates including enterobactin and the siderophore precursor 2,3-dihydroxybenzoic acid [47]. The oxidized forms of these compounds were proposed to bind copper, contributing to Cu elimination [48]. CueO is exported to the periplasm by the TAT system. This suggests that the protein is folded and metallated in the cytoplasm before its exportation to the periplasm. The cytoplasmic metallation of this protein has been proposed to contribute to copper tolerance under anaerobic conditions when the oxidase activity of the protein is inhibited [49].

In addition to the above-mentioned mechanisms, many bacteria employ metallothioneins, polyphosphate and glutathione among others for achieving intracellular or extracellular sequestration in an attempt to reduce reactivity of the Cu ions [50–52].

1.5 Copper Monitoring by Dedicated Sensors

A set of dedicated sensory systems continuously monitor the Cu-quota at the different bacterial compartments and orchestrate the proper response to metal stress [53]. In the cytoplasm, both metal sensing and transcriptional regulation are attained by regulators able to detect the metal and to control the expression of specific target genes. In contrast, periplasmic copper levels are detected by two-component sensors that control the activation state of associated response regulators through phosphorylation.

One of the best characterized sensors of the first group, widely distributed among Gram negative bacteria, belongs to the MerR family of regulators [54]. These proteins, named CueR in *E. coli*, detect either Cu(I), Ag(I) or Au(I) in the cytoplasm using two conserved cysteine residues present in a C-terminus loop [54]. They recognize their target operator sequences by an N-terminal helix-turn-helix motif [55]. Structural data available on *E. coli* CueR revealed that the metal is coordinated in a linear array similar to CopZ-like chaperones [56]. The sequence of the metal binding loop influences its selectivity among metal ion with +1 charge. The single replacement of the amino acid residues at position 113 and 118 for those present in GolS, a CueR-like homologue present in *Salmonella* with an increased selectivity for Au(I), decreased the sensitivity of the sensor to copper while conserved the sensitivity to Au ions [57]. On the other hand, the presence of a single serine residue at the beginning of the dimerization α5-helix together with the hydrophobic environment is essential to excluded metal ions with +2 charge from the binding site (our unpublished results). In the presence of copper, CueR induces the expression of a set of genes that varies depending on the bacterial species [58]. The most prevalent gene controlled by this transcriptional factor encodes for P_{1B}-ATPases. Other target includes genes encoding chaperone, multicopper oxidases and/or other Cu-binding proteins such as CueP [43, 49, 59, 60].

Repressors undergoing intracellular Cu sensing are more common in Gram positive and acid-fast bacteria [53]. Similar to the CueR-like activators, they detect cytoplasmic Cu(I) and regulate the expression of factors required for Cu-resistance, including P_{1B}-ATPases and Cu chaperons [61–64]. In contrast to the activators, the interaction of Cu(I) to the regulator provokes their dissociation from the DNA, leading to the induction of transcription as the result of the derepression. CsoY from *E. hirae* was the first member of the family to be identified [62], and afterwards, other homologues were detected in Gram positive species [61, 63–65]. The active repressor form of CopY is an homodimer with zinc bound. When copper levels increased, the cytoplasmic chaperone CopZ transfers Cu(I) to the conserved

CXCXXXCXC sequence [66], displacing the bound zinc by copper, inactivating the repressor which is released from the DNA allowing transcription of the target genes.

CsoR, first identified in the pathogen *Mycobacterium tuberculosis*, and lately in *Listeria monocytogenes*, *Staphylococcus aureus*, *Corynebacterium glutamicum* and *B. subtilis*, among other Gram positive bacteria, represents another class of Cu(I)-binding repressors [64, 67–70]. As CsoY, CsoR controls the expression of P_{1B}-ATPases and CopZ-like chaperones coding genes, but it does not required a Cu-chaperone to acquired Cu(I). The active form of this repressor is a tetramer. Binding of four Cu(I) ions, one in each monomer, releases the complex from the operator site on the DNA allowing transcription of the gene(s) [71].

The cyanobacterium *Oscillatoria brevis* uses an unusual member of the SmtB/ArsR family of repressors to control intracellular Cu [72]. This protein named BxmR undergoes derepression of its target genes in response to Cu(I), but also to Ag(I), Zn(II), Cd(II) or Pb(II), although the +1 and +2 charge metal ions are bound at different sites in the sensor protein. The Cu(I)-binding site is located at the N-terminal region while Zn(II) is bound at its C-terminus. Interestingly, a Cu(I)-responsive repressor member of the TetR family has been recently described in *E. coli* [20]. This protein named ComR, controls the expression of the divergent gene *comC* in response to copper. ComC localizes to the outer membrane and is proposed to modulate the permeability to copper ions.

Two-component regulatory systems usually detect signals in the periplasm or the membrane to exert transcriptional regulation. Among those involved in copper resistance, the best characterized are the plasmid encoded CopRS and PcoRS from *P. syringae* and some *E. coli* strains, respectively [73, 74], and CusRS from *E. coli* that controls the expression of *cusCFBA* [75]. In these systems, CopS, PcoS and CusS are the membrane-bound sensor components and CopR, PcoR and CusR the corresponding response regulators. As with other members of this family, after Cu (I) detection, presumably in the periplasm, the sensor undergoes an autophosphorylation of a specific histidine [1]. This phosphate is then transferred to a conserved aspartate residue of the response regulator. Phosphorylation of the latter usually increases its DNA affinity, inducing the expression of Cu-resistance determinants to deal with periplasmic copper excess [41].

1.6 Copper and the Innate Immune Response to Control Infections

The innate immune response represents the first line of defense against infections. It involves a group of cells and mechanisms employed to recognize and respond to pathogens in a non-specific manner. Macrophages are an essential part of this response [76]. After detection and uptake of potentially harmful bacteria, a cascade of events is induced, triggering the production of different antimicrobial compounds and the secretion of proinflammatory mediators that serve not only to eliminate the invading microorganism but also to exacerbate the response. This includes an abrupt drop in pH, the generation of reactive nitric oxide and oxygen-derived species, as well as an increase traffic of hydrolytic enzymes and antimicrobial peptides into the lumen of the phagolysosome [77, 78]. Recent reports provide compelling evidence that Cu also plays a role in these events [79, 80].

Copper trafficking in mammalians cells depends on shuttle proteins that deliver the metal to its final destination. One of these proteins is CTR1, a homotrimeric transporter found at the plasma membrane and also in intracellular vesicles [81]. The distribution of CTR1 between these two compartments and the extent to which copper influences this location is cell-specific and is modified depending on the metabolic condition. For example, large amounts of CTR1 protein and the consequent increase in copper uptake are observed in macrophages treated with gamma-interferon or in response to hypoxia, as well as after infection with *M. tuberculosis* [79, 80] (Fig. 1.1). In the same conditions there is an increase in the levels of the copper-transporting ATPases ATP7A in trans-Golgi network that usually supplies the metal to a group of secreted cuproenzymes. When cytoplasmic copper levels increase, binding of the metal to cytoplasmic regions of ATP7A stimulates its traffic to post-Golgi compartments, including cytoplasmic vesicles and the plasma membrane, which allows the rapid elimination of the toxic metal by vesicle-mediated excretion [82, 83]. The treatment of macrophages with proinflammatory agents such as IFN-gamma or lipopolysaccharide (LPS) results in the accumulation of ATP7A in the membrane of endocytic vesicles [84]. The location of ATP7A in these vesicles was proposed to be linked with the increase in CTR1 levels that direct the influx of Cu into the cytoplasm after macrophage activation. Accordingly, it was reported that silencing of ATP7A expression attenuates bacterial killing, suggesting that copper transport into these vesicles is essential for the bactericidal activity of macrophages [84]. In fact, an *E. coli* mutant deleted in *copA* is hypersensitive to killing by macrophages, and this phenotype depends on ATP7A expression [84]. These observations indicate that the controlled accumulation of Cu in specific subcellular compartments is part of the arsenal the host employs as a defense strategy against invading pathogens.

1.7 Copper Resistance and Bacterial Pathogenesis

1.7.1 Salmonella

S. enterica serovar Typhimurium progresses through diverse environments during its infection cycle that are expected to contain variable amounts of copper [85]. Within a mammalian host, it must be able to survive in the gastrointestinal tract, cross the intestinal epithelial barrier and persist within macrophages inside the *Salmonella*-containing vacuole (SCV) [86].

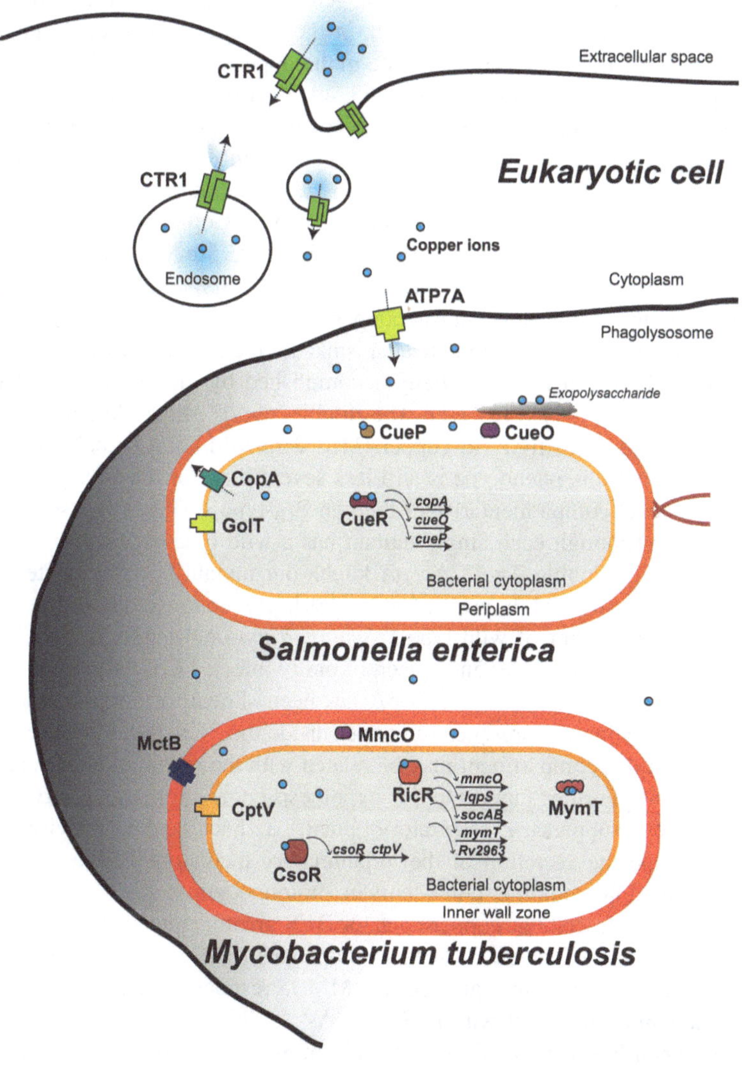

Fig. 1.1 Copper and bacterial pathogenesis. Cu accumulation inside phagolysosome restricts survival and replication of intracellular pathogens during infection. The coordinated action of the host's transporters CTR1 and ATP7A directs the influx of Cu to the bacteria-containing vesicles, which results in Cu accumulation at this compartment. This rise in the copper level is rapidly detected by pathogen-dedicated sensors—CueR in *Salmonella*, and CsoR and RicR in *Mycobacterium*—which induce the expression of a consortium of copper-resistance mechanisms to counteract Cu toxicity and ensure survival. The main Cu-resistance systems described in *Salmonella* and *Mycobacterium* are shown

To cope with copper excess, this pathogen harbors the ancestral copper-resistance regulon controlled by CueR which, in the presence of Cu ions, induces the expression of the P_{1B}-ATPase CopA, the multicopper oxidase CueO (alias CuiD), and the periplasmic CueP coding genes [43, 49, 87] (Fig. 1.1). In aerobiosis, CueO is largely the most important resistance factor with a minimal inhibitory concentration (MIC) of 1.25 mM $CuSO_4$ compared to 4.5 mM for the Δ*copA* strain and 5.5 mM for the wild-type strain [49]. This is well in contrast to the role of these factors in the non-pathogenic *E. coli*, in which both CopA and CueO contribute similarly to copper resistance with MICs of 2.5 and 2.75 mM $CuSO_4$, respectively [49, 88]. These differences probably envisage that *Salmonella* requires a tight protection of the periplasm against copper toxicity, or predict the presence of alternative copper detoxification mechanisms. In fact, in the absence of CopA, resistance to the metal ion is partially accomplished by the GolS-controlled P_{1B}-type ATPase GolT [49] (Fig. 1.1). The double Δ*golT* Δ*copA* knock-out mutant show an increased sensitivity to copper (MIC 2.75 mM $CuSO_4$) compared to the Δ*copA* strain, but this phenotype is still less severe than for the Δ*cueO* mutant.

The functional complementarity of the two P_{1B}-type ATPases is also evident in pathogenesis. Although each single mutant has a wild type survival phenotype in macrophages, the double Δ*golT* Δ*copA* knock-out mutant is outcompeted by the wild-type strain after 24 h post-infection [89]. By contrast, no difference was observed in the number of wild-type *S. typhimurium* or the Δ*copA* Δ*golT* double mutant isolated from the liver and spleen of orally infected C57/BL6 mice. Besides these discrepancies, expression of CopA has been shown to increase inside macrophages, suggesting the presence of copper inside the SCV [89, 90]. The increase in Cu levels in this compartment may be related with the recruitment of ATP7A, as discussed above. Further support to this hypothesis comes from the observation that Cu deficiency suppresses respiratory burst, a bactericidal activity against *Salmonella,* and the secretion of the inflammatory mediators TNF-a, IL-1b, IL-6 and PGE2 in differentiating U937 human promonocytic cells [91]. In fact, the addition of the extracellular copper chelator BCS greatly enhanced *S. typhimurium* survival within bone marrow derived macrophages (BMM), and copper transport genes were dramatically up-regulated in BMM in response to either infection with *S. typhimurium* or treatment with LPS [92]. As we discuss later, the Cu transporter CopA has been linked to virulence not only in *Salmonella* but in other species as well.

The multicopper oxidase CueO has also been involved in *Salmonella* pathogenesis (Fig. 1.1). It has been reported that a mutant deleted in CueO is not only attenuated for virulence in mice, but also exhibited a significantly decreased colonization of liver and spleen [3]. However, its survival is not affected in macrophages in vitro assays. This differs from the enhanced bladder colonization observed in mice for an uropathogenic *E. coli cueO* mutant [93]. The survival advantage exhibited by the mutant in this tissue was not linked to Cu resistance. Instead this strain was shown to have an increased iron acquisition capability compared with the wild-type strain that ensure survival in the iron-limited environment of the mice bladder. Further investigation is required to unravel the balance between copper and iron in particular during host-pathogen interactions.

In a *cueO* mutant strain, copper activates the expression of genes controlled by the multicomponent phosphorelay system *RcsCBA-RcsF*, a regulatory system involved in *Salmonella* virulence [94]. As a consequence, the synthesis of colanic acid is induced and colonies acquired a mucoid aspect (Fig. 1.1). Expression of CueP in this strain suppresses the *cueO* mutant mucoid phenotype, suggesting a role for this protein in alleviating the toxic effect of Cu probably by counteracting the mechanism that activates Rcs in this conditions [94]. One attractive hypothesis is that the induction of the Rcs system and probably the location of the colanic acid exopolysaccharide in the bacterial surface could confer additional protection for bacterial cells living in copper rich environments [95, 96].

Besides the above mechanisms, the *scsABCD* locus also contributes to copper tolerance in *Salmonella* [97]. These genes code for proteins with homology to oxidoreductases of the thioredoxin family that has been proposed to assist to control the envelope oxidative stress working together with the cytoplasmic thioredoxin TrxA [98]. Despite this function, mutants in the *scsABCD* locus have no effect in intracellular replication in RAW264.7 cells, or in mice spleen and liver colonization.

Additionally to the copper resistance mechanisms describe above, *Salmonella* harbors two periplasmic copper-zinc superoxide dismutase proteins (SodCI and SodCII), which catalyze the dismutation of superoxide into oxygen and hydrogen peroxide [99]. The *sodCI* mutant strain has increased susceptibility to be killed by activated murine macrophages as the result of both respiratory burst and nitric oxide production [100]. This mutant is highly attenuated for virulence in a systemic model using C3H/HeN and DBA2 mice [101]. By contrast, the role of the periplasmic superoxide dismutase SodCII in virulence is less clear. Recently it was shown that under copper limitation, CueP is required to metallate SodCII in a pathway that would also involve the transport of Cu(I) from the cytoplasm to the periplasm through CopA or GolT [35].

All together these evidences indicate a clear role of copper in the intracellular life of *Salmonella*. However, more work is required to clarify the individual contribution of the bacterial-copper determinants to *Salmonella* virulence.

1.7.2 Mycobacterium

M. tuberculosis (Mtb) and *M. avium* are intracellular pathogens that primarily infect mononuclear phagocytes [102]. Chronic infection is characterized by the formation of granulomas [103]. Bacteria can persist within granulomas that contained the bacterial spreading. Breakdown of granulomas in the lung promotes release of bacteria and is enhanced by the destruction of lung tissue, which is mediated by the same immune cells necessary for protection during the earlier stages of infection [102]. Once inside macrophages and monocytes, *Mycobacterium* resides within cytoplasmic vacuoles that neither acidify nor fuse with lysosomes [104]. During the course of the infection from 1 to 24 h, copper has been reported to fluctuate from 28.3 to 17.3 μM or 426 to 24.7 μM inside the phagosome of macrophage infected with *M. avium* or *M. tuberculosis*, respectively [105]. Evidence of the role of *Mycobacterium* copper-resistance in pathogenicity comes from studies carried out in guinea pigs. In these animals, larger portions of lung lobes are damaged and the granulomatous responses are more severe when infected with wild type *M. tuberculosis* than with a mutant strain deleted in the Cu P_{1B}-type ATPase CtpV [106]. Moreover, compared with unaffected lung parenchyma, Cu accumulates in isolated primary granulomas, the *Mycobacterium* infection foci [80].

Two related transcriptional regulators, CsoR and RicR, control mycobacterial copper resistance [68, 107] (Fig. 1.1). In the absence of copper, CsoR represses the expression of the *cso* operon, coding for the regulator itself and the P-type ATPase CtpV [68, 107]. The paralogous repressor RicR controls the transcription of five-loci including genes coding for a lipoprotein (LqpS), a Cu(I) binding metallothionein (MymT) that can bind up to 6 Cu atoms per protein, a membrane bound multicopper oxidase (Rv0846c or MmcO), a putative permease (Rv2963), and the *socAB* operon of unknown function [68, 107]. Both MymT and MmcO were reported to affect copper resistance [4, 50], but it remains to be determined the role, if any, of the other members of the RicR regulon in *Mycobacterium* virulence. Preliminary observations indicated that the deficiency of MymT does not impair *M. tuberculosis* virulence in mice [50].

Recently, an outer membrane channel protein named MctB was reported to be essential for both copper resistance and virulence [80] (Fig. 1.1). The burden of *mctB* mutants that accumulate in lungs of infected guinea pigs is reduced more than 1000 fold when compared with those infected with wild type *M. tuberculosis*. The Δ*mctB* mutant is also significantly impaired in dissemination from the lung to the draining lymph nodes compared with the wild type strain, but it is not affected in its ability to disseminate from the lung to the spleen [80]. In BALB/c mice model, 10-fold fewer Δ*mctB* than wild type bacteria were counted in lung homogenates after 20 days post-infection. Moreover, the survival of Δ*mctB* strain was severely compromised when mice were fed with copper [80]. In this condition, the mutant exhibited a 100-fold decrease in cells counts compared to the wild type strain. The description of MctB as a factor required for virulence of *Mycobacterium* in mice could explain why the deficiency of the metallothionein MymT does not impair *M. tuberculosis* virulence [50]. In fact, the contribution of MymT to the bacterial Cu-detoxification inside the eukaryotic cell could be masked by the presence of MctB, which continuously pump copper out of the bacterial cytoplasm [50].

1.7.3 Other Pathogens

Copper has been shown to influence the virulence of other bacteria as well. In *P. aeruginosa*, virulence in mice is severely decreased (20 fold) by mutations in the copper exporter CueA [108]. A mutation in the gene coding for this pump also compromises the fitness of *P. fluorescence* in plant rhizosphere [109]. In *Legionella pneumophila*, the expression of genes coding for two putative copper efflux transporters, CopA1 and CopA2, are induced upon phagocytosis by macrophages [110]. Nevertheless, deletion of the island containing these copper efflux coding genes does not affect survival within macrophages or amoebas.

Mutations in the gene encoding the P_{1B}-type ATPase CtpA of *L. monocytogenes* impair survival in mice tissues, although no differences were observed in intracellular growth in J774 and HeLa cell lines [111]. Interestingly, deletion of CopA, encoding another Cu transporting ATPase (*lmo1872*, different from *ctpA*) does not affect the virulence in mice [70]. It was argued that this bacterium rapidly escapes from eukaryotic vesicles and the metal rise inside the vesicles could not affect bacterial survival. Alternatively, this could be explained because of the redundancy of the copper resistance factors, as in *Salmonella*. Besides *copA* and *ctpA*, there are other loci in *Listeria* coding for heavy-metal transporting ATPases such as *lmo0642* and *cadA* that could also contribute to Cu efflux [112]. Copper resistance is important for *S. pneumoniae* virulence as well. It has been reported that the survival of mice infected intranasally with a *copA mutant* is severely impaired when compared to that infected with the wild type D39 strain [63]. Moreover, the growth of pneumococci *copA* mutants in the nasopharynx is also compromised at 12, 24 and 36 h post infection. Further studies combining mutants in the copper resistance factors would be necessary to clarify the copper-mediated killing of *Listeria* and *Streptococcus* in a host model.

1.8 Concluding Remarks

The role of Cu as an external microbicide is well known, as it is its capability to control infections in plants and animals. In fact, Cu has been used in healthcare and to improve crops for thousands of years [113]. Moreover, patients that have Cu deficiency disorders or animals subjected to a Cu deficient diet are highly prone to microbial infections [114]. The evidences accumulated in the last years indicated that copper can restrict bacterial growth during infection cycle, particularly of pathogens like *Salmonella* and *Mycobacterium* that undergo intracellular survival and replication cycles inside phagolysosomal compartments. However, more basic research is still needed to understand the direct function of copper against these and other pathogens. For example, the transcriptional profile of *Salmonella* in the presence of copper will uncover the whole set of copper-regulated genes involved in virulence. Additionally, it urges to determine the fluctuation in copper levels inside the *Salmonella*-containing vacuole (SCV), and to analyze the effects of combining different mutants in genes belonging to CueR and GolS regulon on both cellular and systemic virulence phenotypes. In *Mycobacterium*, there are still many copper regulated genes not characterized or tested in virulence. Furthermore, the results already obtained with mutants in copper transporters genes promise a fertile future for this field. These findings will also open novel fields in antimicrobial compound research. For example, the membrane-permeable bis-thiosemicarbazones ATSM and GTSM were tested in vitro as potential anti-*M. tuberculosis* drugs that exploit the bactericidal properties of copper ions [115]. Some copper binding drugs are already used to treat alcohol dependence (disulfiram) and Wilson's disease (penicillamine) [116], and could be adapted to infection treatments. Also, innovative research is focus on copper binding as anticancer and anti-HIV drugs, and this could be extended to novel antibiotic development.

References

1. Rensing C, Grass G (2003) *Escherichia coli* mechanisms of copper homeostasis in a changing environment. FEMS Microbiol Rev 27:197–213
2. Kim EH, Rensing C, McEvoy MM (2010) Chaperone-mediated copper handling in the periplasm. Nat Prod Rep 27:711–719
3. Achard ME et al (2010) The multi-copper-ion oxidase CueO of *Salmonella enterica* serovar Typhimurium is required for systemic virulence. Infect Immun 78:2312–2319
4. Rowland JL, Niederweis M (2012) Resistance mechanisms of *Mycobacterium tuberculosis* against phagosomal copper overload. Tuberculosis (Edinb) 92:202–210
5. MacPherson IS, Murphy ME (2007) Type-2 copper-containing enzymes. Cell Mol Life Sci 64:2887–2899
6. Rosenzweig AC, Sazinsky MH (2006) Structural insights into dioxygen-activating copper enzymes. Curr Opin Struct Biol 16:729–735
7. Choi M, Davidson VL (2011) Cupredoxins–a study of how proteins may evolve to use metals for bioenergetic processes. Metallomics 3:140–151
8. Eskici G, Axelsen PH (2012) Copper and oxidative stress in the pathogenesis of Alzheimer's disease. Biochemistry 51:6289–6311
9. Jomova K, Vondrakova D, Lawson M, Valko M (2010) Metals, oxidative stress and neurodegenerative disorders. Mol Cell Biochem 345:91–104
10. Lalioti V, Muruais G, Tsuchiya Y, Pulido D, Sandoval IV (2009) Molecular mechanisms of copper homeostasis Front Biosci (Landmark. Ed) 14:4878–4903
11. Rubino JT, Franz KJ (2012) Coordination chemistry of copper proteins: how nature handles a toxic cargo for essential function. J Inorg Biochem 107:129–143
12. Liochev SI, Fridovich I (2002) The Haber-Weiss cycle—70 years later: an alternative view. Redox Rep 7:55–57
13. Macomber L, Imlay JA (2009) The iron-sulfur clusters of dehydratases are primary intracellular targets of copper toxicity. Proc Natl Acad Sci USA 106:8344–8349
14. Fung DK, Lau WY, Chan WT, Yan A (2013) Copper efflux is induced during anaerobic amino acid limitation in *Escherichia coli* to protect iron-sulfur cluster enzymes and biogenesis. J Bacteriol 195:4556–4568
15. Rensing C, McDevitt SF (2013) The copper metallome in prokaryotic cells. Met Ions Life Sci 12:417–450
16. Decaria L, Bertini I, Williams RJ (2011) Copper proteomes, phylogenetics and evolution. Metallomics 3:56–60
17. Cha JS, Cooksey DA (1993) Copper hypersensitivity and uptake in pseudomonas syringae containing cloned components of the copper resistance operon. Appl Environ Microbiol 59:1671–1674

18. Chillappagari S, Miethke M, Trip H, Kuipers OP, Marahiel MA (2009) Copper acquisition is mediated by YcnJ and regulated by YcnK and CsoR in *Bacillus subtilis*. J Bacteriol 191:2362–2370
19. Ekici S, Yang H, Koch HG, Daldal F (2012) Novel transporter required for biogenesis of cbb3-type cytochrome c oxidase in *Rhodobacter capsulatus*. MBio 3
20. Mermod M, Magnani D, Solioz M, Stoyanov JV (2012) The copper-inducible ComR (YcfQ) repressor regulates expression of ComC (YcfR), which affects copper permeability of the outer membrane of *Escherichia coli*. Biometals 25:33–43
21. Grass G et al (2005) The metal permease ZupT from *Escherichia coli* is a transporter with a broad substrate spectrum. J Bacteriol 187:1604–1611
22. Janulczyk R, Pallon J, Bjorck L (1999) Identification and characterization of a *Streptococcus pyogenes* ABC transporter with multiple specificity for metal cations. Mol Microbiol 34:596–606
23. Lewinson O, Lee AT, Rees DC (2009) A P-type ATPase importer that discriminates between essential and toxic transition metals. Proc Natl Acad Sci USA 106:4677–4682
24. Kenney GE, Rosenzweig AC (2012) Chemistry and biology of the copper chelator methanobactin. ACS Chem Biol 7:260–268
25. Singleton C, Le Brun NE (2007) Atx1-like chaperones and their cognate P-type ATPases: copper-binding and transfer. Biometals 20:275–289
26. Gonzalez-Guerrero M, Hong D, Arguello JM (2009) Chaperone-mediated Cu+ delivery to Cu+ transport ATPases: requirement of nucleotide binding. J Biol Chem 284:20804–20811
27. Banci L, Bertini I, McGreevy KS, Rosato A (2010) Molecular recognition in copper trafficking. Nat Prod Rep 27:695–710
28. Tottey S, Rondet SA, Borrelly GP, Robinson PJ, Rich PR, Robinson NJ (2002) A copper metallochaperone for photosynthesis and respiration reveals metal-specific targets, interaction with an importer, and alternative sites for copper acquisition. J Biol Chem 277:5490–5497
29. Fu Y et al (2013) A new structural paradigm in copper resistance in *Streptococcus pneumoniae*. Nat Chem Biol 9:177–183
30. Xue Y et al (2008) Cu(I) recognition via cation-pi and methionine interactions in CusF. Nat Chem Biol 4:107–109
31. Loftin IR et al (2005) A novel copper-binding fold for the periplasmic copper resistance protein CusF. Biochemistry 44:10533–10540
32. Mealman TD et al (2012) N-terminal region of CusB is sufficient for metal binding and metal transfer with the metallochaperone CusF. Biochemistry 51:6767–6775
33. Thompson AK, Gray J, Liu A, Hosler JP (2012) The roles of *Rhodobacter sphaeroides* copper chaperones PCu(A)C and Sco (PrrC) in the assembly of the copper centers of the aa (3)-type and the cbb(3)-type cytochrome c oxidases. Biochim Biophys Acta 1817:955–964
34. Balatri E, Banci L, Bertini I, Cantini F, Ciofi-Baffoni S (2003) Solution structure of Sco1: a thioredoxin-like protein Involved in cytochrome c oxidase assembly. Structure 11:1431–1443
35. Osman D et al (2013) The copper supply pathway to a Salmonella Cu, Zn-superoxide dismutase (SodCII) involves P(1B)-type ATPase copper efflux and periplasmic CueP. Mol Microbiol 87:466–477
36. Yoon BY et al (2013) Structure of the periplasmic copper-binding protein CueP from *Salmonella enterica* serovar Typhimurium. Acta Crystallogr D Biol Crystallogr 69:1867–1875
37. Rosenzweig AC, Arguello JM (2012) Toward a molecular understanding of metal transport by P(1B)-type ATPase. Curr Top Membr 69:113–136
38. Padilla-Benavides T, McCann CJ, Arguello JM (2013) The mechanism of Cu+ transport ATPases: interaction with CU+ chaperones and the role of transient metal-binding sites. J Biol Chem 288:69–78
39. Raimunda D, Gonzalez-Guerrero M, Leeber BW III, Arguello JM (2011) The transport mechanism of bacterial Cu+-ATPases: distinct efflux rates adapted to different function. Biometals 24:467–475

40. Franke S, Grass G, Rensing C, Nies DH (2003) Molecular analysis of the copper-transporting efflux system CusCFBA of *Escherichia coli*. J Bacteriol 185:3804–3812
41. Mealman TD, Blackburn NJ, McEvoy MM (2012) Metal export by CusCFBA, the periplasmic Cu(I)/Ag(I) transport system of *Escherichia coli*. Curr Top Membr 69:163–196
42. Su CC, Long F, Zimmermann MT, Rajashankar KR, Jernigan RL, Yu EW (2011) Crystal structure of the CusBA heavy-metal efflux complex of *Escherichia coli*. Nature 470:558–562
43. Pontel LB, Soncini FC (2009) Alternative periplasmic copper-resistance mechanisms in Gram negative bacteria. Mol Microbiol 73:212–225
44. Grass G, Rensing C (2001) CueO is a multi-copper oxidase that confers copper tolerance in *Escherichia coli*. Biochem Biophys Res Commun 286:902–908
45. Roberts SA et al (2002) Crystal structure and electron transfer kinetics of CueO, a multicopper oxidase required for copper homeostasis in *Escherichia coli*. Proc Natl Acad Sci USA 99:2766–2771
46. Singh SK, Grass G, Rensing C, Montfort WR (2004) Cuprous oxidase activity of CueO from *Escherichia coli*. J Bacteriol 186:7815–7817
47. Kim C, Lorenz WW, Hoopes JT, Dean JF (2001) Oxidation of phenolate siderophores by the multicopper oxidase encoded by the *Escherichia coli yacK* gene. J Bacteriol 183:4866–4875
48. Grass G et al (2004) Linkage between catecholate siderophores and the multicopper oxidase CueO in *Escherichia coli*. J Bacteriol 186:5826–5833
49. Espariz M, Checa SK, Audero ME, Pontel LB, Soncini FC (2007) Dissecting the *Salmonella* response to copper. Microbiology 153:2989–2997
50. Gold B et al (2008) Identification of a copper-binding metallothionein in pathogenic *Mycobacteria*. Nat Chem Biol 4:609–616
51. Potter AJ, Trappetti C, Paton JC (2012) *Streptococcus pneumoniae* uses glutathione to defend against oxidative stress and metal ion toxicity. J Bacteriol 194:6248–6254
52. Remonsellez F, Orell A, Jerez CA (2006) Copper tolerance of the thermoacidophilic archaeon *Sulfolobus metallicus*: possible role of polyphosphate metabolism. Microbiology 152:59–66
53. Reyes-Caballero H, Campanello GC, Giedroc DP (2011) Metalloregulatory proteins: metal selectivity and allosteric switching. Biophys Chem 156:103–114
54. Brown NL, Stoyanov JV, Kidd SP, Hobman JL (2003) The MerR family of transcriptional regulators. FEMS Microbiol Rev 27:145–163
55. Humbert MV, Rasia RM, Checa SK, Soncini FC (2013) Protein signatures that promote operator selectivity among paralog MerR monovalent metal ion regulators. J Biol Chem 288:20510–20519
56. Changela A et al (2003) Molecular basis of metal-ion selectivity and zeptomolar sensitivity by CueR. Science 301:1383–1387
57. Ibanez MM, Cerminati S, Checa SK, Soncini FC (2013) Dissecting the metal selectivity of MerR monovalent metal ion sensors in *Salmonella*. J Bacteriol 195:3084–3092
58. Perez Audero ME, Podoroska BM, Ibanez MM, Cauerhff A, Checa SK, Soncini FC (2010) Target transcription binding sites differentiate two groups of MerR-monovalent metal ion sensors. Mol Microbiol 78:853–865
59. Stoyanov JV, Hobman JL, Brown NL (2001) CueR (YbbI) of Escherichia coli is a MerR family regulator controlling expression of the copper exporter CopA. Mol Microbiol 39: 502–511
60. Outten FW, Outten CE, Hale J, O'Halloran TV (2000) Transcriptional activation of an *Escherichia coli* copper efflux regulon by the chromosomal MerR homologue, cueR. J Biol Chem 275:31024–31029
61. Magnani D, Barre O, Gerber SD, Solioz M (2008) Characterization of the CopR regulon of *Lactococcus lactis* IL1403. J Bacteriol 190:536–545
62. Portmann R, Poulsen KR, Wimmer R, Solioz M (2006) CopY-like copper inducible repressors are putative 'winged helix' proteins. Biometals 19:61–70
63. Shafeeq S et al (2011) The cop operon is required for copper homeostasis and contributes to virulence in *Streptococcus pneumoniae*. Mol Microbiol 81:1255–1270

64. Smaldone GT, Helmann JD (2007) CsoR regulates the copper efflux operon *copZA* in *Bacillus subtilis*. Microbiology 153:4123–4128
65. Cantini F, Banci L, Solioz M (2009) The copper-responsive repressor CopR of *Lactococcus lactis* is a 'winged helix' protein. Biochem J 417:493–499
66. Cobine P, Wickramasinghe WA, Harrison MD, Weber T, Solioz M, Dameron CT (1999) The *Enterococcus hirae* copper chaperone CopZ delivers copper(I) to the CopY repressor. FEBS Lett 445:27–30
67. Teramoto H, Inui M, Yukawa H (2012) *Corynebacterium glutamicum* CsoR acts as a transcriptional repressor of two copper/zinc-inducible P(1B)-type ATPase operons. Biosci Biotechnol Biochem 76:1952–1958
68. Liu T et al (2007) CsoR is a novel *Mycobacterium tuberculosis* copper-sensing transcriptional regulator. Nat Chem Biol 3:60–68
69. Grossoehme N et al (2011) Control of copper resistance and inorganic sulfur metabolism by paralogous regulators in *Staphylococcus aureus*. J Biol Chem 286:13522–13531
70. Corbett D, Schuler S, Glenn S, Andrew PW, Cavet JS, Roberts IS (2011) The combined actions of the copper-responsive repressor CsoR and copper-metallochaperone CopZ modulate CopA-mediated copper efflux in the intracellular pathogen *Listeria monocytogenes*. Mol Microbiol 81:457–472
71. Ma Z, Cowart DM, Scott RA, Giedroc DP (2009) Molecular insights into the metal selectivity of the copper(I)-sensing repressor CsoR from *Bacillus subtilis*. Biochemistry 48:3325–3334
72. Liu T et al (2008) A Cu(I)-sensing ArsR family metal sensor protein with a relaxed metal selectivity profile. Biochemistry 47:10564–10575
73. Brown NL, Barrett SR, Camakaris J, Lee BT, Rouch DA (1995) Molecular genetics and transport analysis of the copper-resistance determinant (pco) from *Escherichia coli* plasmid pRJ1004. Mol Microbiol 17:1153–1166
74. Mills SD, Jasalavich CA, Cooksey DA (1993) A two-component regulatory system required for copper-inducible expression of the copper resistance operon of *Pseudomonas syringae*. J Bacteriol 175:1656–1664
75. Munson GP, Lam DL, Outten FW, O'Halloran TV (2000) Identification of a copper-responsive two-component system on the chromosome of *Escherichia coli* K-12. J Bacteriol 182:5864–5871
76. Smith LM, May RC (2013) Mechanisms of microbial escape from phagocyte killing. Biochem Soc Trans 41:475–490
77. Nathan C, Shiloh MU (2000) Reactive oxygen and nitrogen intermediates in the relationship between mammalian hosts and microbial pathogens. Proc Natl Acad Sci USA 97:8841–8848
78. Shiloh MU, Nathan CF (2000) Reactive nitrogen intermediates and the pathogenesis of *Salmonella* and *mycobacteria*. Curr Opin Microbiol 3:35–42
79. White C et al (2009) Copper transport into the secretory pathway is regulated by oxygen in macrophages. J Cell Sci 122:1315–1321
80. Wolschendorf F et al (2011) Copper resistance is essential for virulence of *Mycobacterium tuberculosis*. Proc Natl Acad Sci USA 108:1621–1626
81. Zhou B, Gitschier J (1997) hCTR1: a human gene for copper uptake identified by complementation in yeast. Proc Natl Acad Sci USA 94:7481–7486
82. Hamza I, Prohaska J, Gitlin JD (2003) Essential role for Atox1 in the copper-mediated intracellular trafficking of the Menkes ATPase. Proc Natl Acad Sci USA 100:1215–1220
83. Petris MJ, Mercer JF, Culvenor JG, Lockhart P, Gleeson PA, Camakaris J (1996) Ligand-regulated transport of the Menkes copper P-type ATPase efflux pump from the Golgi apparatus to the plasma membrane: a novel mechanism of regulated trafficking. EMBO J 15:6084–6095
84. White C, Lee J, Kambe T, Fritsche K, Petris MJ (2009) A role for the ATP7A copper-transporting ATPase in macrophage bactericidal activity. J Biol Chem 284:33949–33956
85. Samanovic MI, Ding C, Thiele DJ, Darwin KH (2012) Copper in microbial pathogenesis: meddling with the metal. Cell Host Microbe 11:106–115

86. Baumler AJ, Winter SE, Thiennimitr P, Casadesus J (2011) Intestinal and chronic infections: *Salmonella* lifestyles in hostile environments. Environ Microbiol Rep 3:508–517
87. Kim JS, Kim MH, Joe MH, Song SS, Lee IS, Choi SY (2002) The *sctR* of *Salmonella enterica* serova Typhimurium encoding a homologue of MerR protein is involved in the copper-responsive regulation of cuiD. FEMS Microbiol Lett 210:99–103
88. Outten FW, Huffman DL, Hale JA, O'Halloran TV (2001) The independent cue and cus systems confer copper tolerance during aerobic and anaerobic growth in *Escherichia coli.* J Biol Chem 276:30670–30677
89. Osman D et al (2010) Copper homeostasis in *Salmonella* is atypical and copper-CueP is a major periplasmic metal complex. J Biol Chem 285:25259–25268
90. Heithoff DM, Conner CP, Hanna PC, Julio SM, Hentschel U, Mahan MJ (1997) Bacterial infection as assessed by in vivo gene expression. Proc Natl Acad Sci USA 94:934–939
91. Huang ZL, Failla ML (2000) Copper deficiency suppresses effector activities of differentiated U937 cells. J Nutr 130:1536–1542
92. Achard ME et al (2012) Copper redistribution in murine macrophages in response to *Salmonella* infection. Biochem J 444:51–57
93. Tree JJ, Ulett GC, Ong CL, Trott DJ, McEwan AG, Schembri MA (2008) Trade-off between iron uptake and protection against oxidative stress: deletion of *cueO* promotes uropathogenic *Escherichia coli* virulence in a mouse model of urinary tract infection. J Bacteriol 190: 6909–6912
94. Pontel LB, Pezza A, Soncini FC (2010) Copper stress targets the rcs system to induce multiaggregative behavior in a copper-sensitive *Salmonella* strain. J Bacteriol 192: 6287–6290
95. Gonzalez AG et al (2010) Adsorption of copper on *Pseudomonas aureofaciens*: protective role of surface exopolysaccharides. J Colloid Interface Sci 350:305–314
96. Yadav KK, Mandal AK, Chakraborty R (2013) Copper susceptibility in *Acinetobacter junii* BB1A is related to the production of extracellular polymeric substances. Antonie Van Leeuwenhoek 104:261–269
97. Gupta SD, Wu HC, Rick PD (1997) A *Salmonella typhimurium* genetic locus which confers copper tolerance on copper-sensitive mutants of *Escherichia coli.* J Bacteriol 179:4977–4984
98. Anwar N, Sem XH, Rhen M (2013) Oxidoreductases that act as conditional virulence suppressors in *Salmonella enterica* serovar Typhimurium. PLoS One 8:e64948
99. Uzzau S, Bossi L, Figueroa-Bossi N (2002) Differential accumulation of *Salmonella*[Cu, Zn] superoxide dismutases SodCI and SodCII in intracellular bacteria: correlation with their relative contribution to pathogenicity. Mol Microbiol 46:147–156
100. De Groote MA et al (1997) Periplasmic superoxide dismutase protects *Salmonella* from products of phagocyte NADPH-oxidase and nitric oxide synthase. Proc Natl Acad Sci USA 94:13997–14001
101. Ammendola S et al (2008) Regulatory and structural differences in the Cu, Zn-superoxide dismutases of Salmonella enterica and their significance for virulence. J Biol Chem 283:13688–13699
102. Repasy T et al (2013) Intracellular bacillary burden reflects a burst size for *Mycobacterium tuberculosis* in vivo. PLoS Pathog 9:e1003190
103. Hunter,R.L. (2011) Pathology of post primary tuberculosis of the lung: an illustrated critical review. Tuberculosis. (Edinb.) 91:497–509
104. Flannagan RS, Cosio G, Grinstein S (2009) Antimicrobial mechanisms of phagocytes and bacterial evasion strategies. Nat Rev Microbiol 7:355–366
105. Wagner D, Maser J, Moric I, Vogt S, Kern WV, Bermudez LE (2006) Elemental analysis of the *Mycobacterium avium* phagosome in Balb/c mouse macrophages. Biochem Biophys Res Commun 344:1346–1351
106. Ward SK, Abomoelak B, Hoye EA, Steinberg H, Talaat AM (2010) CtpV: a putative copper exporter required for full virulence of *Mycobacterium tuberculosis*. Mol Microbiol 77: 1096–1110

107. Festa RA et al (2011) A novel copper-responsive regulon in *Mycobacterium tuberculosis*. Mol Microbiol 79:133–148
108. Schwan WR, Warrener P, Keunz E, Stover CK, Folger KR (2005) Mutations in the cueA gene encoding a copper homeostasis P-type ATPase reduce the pathogenicity of *Pseudomonas aeruginosa* in mice. Int J Med Microbiol 295:237–242
109. Zhang XX, Rainey PB (2007) The role of a P1-type ATPase from *Pseudomonas fluorescens* SBW25 in copper homeostasis and plant colonization. Mol Plant Microbe Interact 20: 581–588
110. Rankin S, Li Z, Isberg RR (2002) Macrophage-induced genes of *Legionella pneumophila*: protection from reactive intermediates and solute imbalance during intracellular growth. Infect Immun 70:3637–3648
111. Francis MS, Thomas CJ (1997) Mutants in the CtpA copper transporting P-type ATPase reduce virulence of *Listeria monocytogenes*. Microb Pathog 22:67–78
112. Renier S, Micheau P, Talon R, Hebraud M, Desvaux M (2012) Subcellular localization of extracytoplasmic proteins in monoderm bacteria: rational secretomics-based strategy for genomic and proteomic analyses. PLoS One 7:e42982
113. Borkow G, Gabbay J (2005) Copper as a biocidal tool. Curr Med Chem 12:2163–2175
114. Munoz C, Rios E, Olivos J, Brunser O, Olivares M (2007) Iron, copper and immunocompetence. Br J Nutr 98(Suppl 1):S24–S28
115. Speer A et al (2013) Copper-boosting compounds: a novel concept for antimycobacterial drug discovery. Antimicrob Agents Chemother 57:1089–1091
116. Wadhwa S, Mumper RJ (2013) D-penicillamine and other low molecular weight thiols: review of anticancer effects and related mechanisms. Cancer Lett 337:8–21

Chapter 2
Shewanella oneidensis and Extracellular Electron Transfer to Metal Oxides

Daad Saffarini, Ken Brockman, Alex Beliaev, Rachida Bouhenni and Sheetal Shirodkar

Abstract Anaerobic metal reduction by bacteria plays an important role in biogeochemical cycles, bioremediation, and in biotechnological applications such as electricity generation. *Shewanella oneidensis* is one of the best-studied metal reducing bacteria and its analysis led to the identification of the mechanisms this bacterium uses for respiratory metal reduction. The major proteins involved in metal reduction in *S. oneidensis* consist of an outer membrane decaheme *c*-type cytochrome (MtrC), an outer membrane porin (MtrB) and a periplasmic decaheme *c*-type cytochrome (MtrA). These proteins form a complex that is located on the outer cell surface and transfers electrons extracellularly to the metal oxides. Although other proteins, such as the outer membrane decaheme *c*-type cytochrome OmcA, are thought to be involved in metal reduction, their role in this process appears to be minor. Several mechanisms to explain the extracellular electron transfer to metal oxides have been proposed. These include direct contact of cells with metal oxides, the use of flavins or electron shuttles, and the use of conductive appendages or nanowires. Flavins, which are thought to allow metal reduction at a

D. Saffarini (✉)
Department of Biological Sciences, University of Wisconsin-Milwaukee, Milwaukee, WI 53211, USA
e-mail: daads@uwm.edu

K. Brockman
The Research Institute, Nationwide Children's Hospital, The Ohio State University College of Medicine, Columbus, OH 43205, USA
e-mail: kenneth.brockman@nationwidechildrens.org

A. Beliaev
Pacific Northwest National Laboratory, Richland, WA, USA
e-mail: Alex.Beliaev@pnnl.gov

R. Bouhenni
Summa Health System, Division of Opthalmology Research, Akron, OH 44309, USA
e-mail: bouhennir@summa-health.org

S. Shirodkar
Amity Institute of Biotechnology, Uttar Pradesh, India
e-mail: sshirodkar@amity.edu

D. Saffarini (ed.), *Bacteria-Metal Interactions*,
DOI 10.1007/978-3-319-18570-5_2

distance from the cells, have been shown to function as cofactors that bind to the outer membrane cytochromes and mediate electron transfer. Conductive appendages or pili, also known as nanowires, have been implicated in mediating electron transfer at a distance. However, *S. oneidensis* mutants that lack pili are able to reduce metals similar to the wild type. Recently, these appendages have been shown to consist of membrane extensions and membrane vesicles. Thus, metal reduction by *S. oneidensis* appears to be mostly the result of direct contact of cell's outer membrane cytochromes with the insoluble metal oxides.

Keywords Metal reduction · *Shewanella oneidensis* · Extracellular electron transfer · Electron shuttles · Nanowires · MtrC · MtrA · MtrB

2.1 Introduction

Iron is an essential micronutrient for almost all living organisms and is one of the most abundant elements on earth. In nature, iron exists in either reduced (Fe^{+2}) or oxidized (Fe^{+3}) forms with speciation determined by key environmental variables such as dissolved oxygen tension and pH. Microorganisms are major contributors to cycling of iron between the oxidized and reduced forms, a process that has become known as the microbial "ferrous wheel" (see [31, 49, 103] and references within). Bacteria and Archaea can use reduced iron as an electron source in aerobic, anaerobic, and acidic environments. As a result, Fe(II) is oxidized to Fe(III), which can then be used by metal reducing bacteria as a terminal electron acceptor for anaerobic respiration. The two best-studied metal reducing bacteria, *Shewanella oneidensis* and *Geobacter metallireducens*, were almost simultaneously isolated in pure culture in 1988 [54, 70]. Since then, many other metal reducing bacterial and archaeal species capable of respiratory metal reduction have been isolated and identified. These include facultative anaerobic bacteria such *S. putrefaciens*, *S. loihica*, *Pantoea agglomerans*, and *Thermus* strain SA-01 [35, 36, 48, 76], anaerobic bacteria such as *G. metallireducens* (reviewed in [53]) and *Ferribacterium limneticum* [22, 23], and anaerobic archaea such as *Geoglobus ahangari* [47]. Since the isolation of these organisms, our understanding of the molecular mechanisms of metal reduction, its involvement in biogeochemical cycles, and its potential use in bioremediation and electricity production, has increased exponentially. In this chapter, we focus on *S. oneidensis* MR-1 and discuss the molecular mechanisms this bacterium uses to transfer electrons extracellularly to metal oxides.

2.2 The *Shewanella* Genus

Members of the genus *Shewanella* are Gram-negative γ-Proteobacteria. They are widespread in diverse environments that include freshwater and marine sediments and water columns, crude oil pipelines, hydrothermal vents, iron-rich microbial mats, activated sludge, and marine fish ([36, 93, 113] for review). Although some *Shewanella* species were recovered from freshwater environments, these isolates are thought to be of marine origin and their presence in freshwater systems is predicted to be recent [43]. The *Shewanella* genus is best known for extracellular electron transfer and, with the exception of *S. denitrificans*, all species sequenced to date have the genes required for this process. The DOE Joint Genome Institute site (http://img.jgi.doe.gov/cgi-bin/w/main.cgi) includes the genome sequences of 36 *Shewanella* species, 23 of which are complete. Analysis of these genomes provided insight into the environmental adaptation and evolution of the *Shewanella* species and revealed diverse metabolic abilities among its members [36].

2.2.1 Shewanella oneidensis *MR-1*

S. oneidensis MR-1 is one of the best characterized members of the *Shewanella* genus and the first to have its genome sequenced [44]. It was initially isolated as a Mn(IV) reducer from Oneida Lake sediments and identified as *Altermonas putrefaciens* [70] before being classified as *Shewanella oneidensis* [106]. *S. oneidensis* MR-1 uses fermentation products as carbon and/or energy sources and has a well-developed chitin utilization system [76, 114]. Metabolically, it is the most diverse of the *Shewanella* species with regard to the electron acceptors it can use for respiration which include O_2, fumarate, NO_3^-, NO_2^-, trimethylamine N-oxide (TMAO), dimethylsulfoxide (DMSO), iron and manganese oxides, and sulfur species such as elemental sulfur and sulfite [25, 36, 67, 70, 76, 102]. Radionuclides and toxic metals such as Tc, U, Cr, can also serve as electron acceptors [6, 52, 59, 60, 73, 75, 76]. Forty-one *c* cytochromes are encoded by the *S. oneidensis* MR-1 genome [44, 89, 104], reflecting its ability to use a wide array of electron acceptors.

The central metal reductase complex in *S. oneidensis* MR-1 is composed of three subunits, MtrB, MtrC, and MtrA. These proteins are encoded by the *mtrCAB* operon that is expressed under microaerobic and anaerobic conditions, even in the absence of metal electron acceptors. Contrary to expectations, the expression of *mtrCAB* is highest in the presence of fumarate and not in the presence of metal oxides [4, 5]. Although these genes are required for metal reduction, their expression is decreased when fumarate-grown cells are transferred to media containing metal oxides [5]. Elevated gene expression under fumarate-growth conditions was also observed for *cymA* and *omcA* [4, 5] that have roles in metal reduction as described in more detail below. Expression of *mtrCAB* and *omcA* is controlled by the cAMP receptor protein CRP [17, 92]. This protein regulates the expression of many anaerobic reductase genes in *S. oneidensis* MR-1, unlike its role in the regulation of carbon metabolism in other bacteria. The role of CRP in anaerobic respiration is not limited to *S. oneidensis* MR-1. A similar mode of regulation has been shown in *Shewanella* sp ANA-3 [68] suggesting that this may be a property of the *Shewanella* genus.

2.3 The *S. oneidensis* MR-1 Metal Reduction Electron Transport Chain

2.3.1 CymA Links the Metal Reductase to the Menaquinol Pool

CymA is a 21 kDa membrane-anchored *c*-type cytochrome that belongs to the NapC/NirT family of quinol dehydrogenases [74, 97]. In contrast to its family members, CymA lacks specificity and is involved in electron transfer to multiple terminal reductases [69, 74]. CymA-dependent reductases, which include the nitrate, nitrite, DMSO, Fe(III), and fumarate reductases, appear to be located in the periplasm or outer membrane of *S. oneidensis* MR-1 [21, 37, 40, 41, 51, 66, 72, 74, 96]. In contrast, CymA is not involved in electron transfer to inner membrane-anchored enzymes such as the TMAO, thiosulfate, and sulfite reductases [8, 16, 25, 38, 102].

CymA is tetraheme *c* cytochrome that binds one high-spin and three low-spin hemes [58]. It is a menaquinol-7 dehydrogenase and its activity is inhibited by the respiratory chain inhibitor 2-n-heptyl-4-hydroxyquinoline-N-oxide (HQNO) and by site directed mutagenesis of the putative quinol-binding site [26, 63, 74, 116]. Interestingly, increased expression of SirCD, which is predicted to function as a quinol oxidoreductase during sulfite reduction in *S. oneidensis* [102], can complement an *S. oneidensis* MR-1 *cymA* mutant and allow reduction of Fe(III) and other electron acceptors [19].

Notably, the location of CymA in the inner membrane does not appear to be essential for its interaction with the menaquinol pool. A soluble CymA that lacks the membrane-spanning domain (the first 35 amino acids of the protein) can complement a *cymA* mutant [96]. Because CymA is thought to interact with menaquinones in the cytoplasmic membrane, it is not clear how it can still function in the absence of its membrane anchor. Zagar and Saltikov suggested that additional sites in CymA could interact with the inner membrane and allow further interactions with the quinol pool [116].

2.3.2 MtrA: A Periplasmic Decaheme c Cytochrome

MtrA is a 32 kDa decaheme *c*-type cytochrome that has been shown to be essential for metal reduction in *S. oneidensis* MR-1 and *Shewanella* sp. ANA-3 [3, 87]. The sequence of MtrA includes a leader peptide and its secretion into the periplasmic space was confirmed by heme staining and Western blot analysis using MtrA-specific antibodies [3, 87]. In cells lysed by osmotic shock, the protein is present mostly in the outer membrane and is associated with MtrCB with a 1:1:1 stoicheometry [13, 91]. MtrA binds 10 low-spin hemes, and has a low amino acid to heme ratio compared to other heme-containing proteins [32, 84]. Based on small-angle X-ray scattering and analytical centrifugation data, MtrA is estimated to be a monomeric protein of 104 Å in length [32]. Using a bacterial two-hybrid system, Borloo et al. [9] determined that MtrA interacts with CymA supporting the hypothesis that CymA transfers electrons directly to the terminal metal reductase. This interaction, however, appears to be transient and CymA is able to reduce MtrA in vitro without the formation of a CymA-MtrC complex [33]. In addition to its role in electron transfer, MtrA appears to be required for stability of the outer membrane porin MtrB [94].

2.3.3 MtrB: An Outer Membrane Porin

MtrB is an outer membrane protein that is essential for metal reduction [2, 20] but is the least studied of the metal reductase components. Computer analysis using PRED-TMBB and proteinase K digestion of MtrB-containing proteoliposomes predicted MtrB to have 28 β-strands that form the transmembrane β-barrel, periplasmic N-terminus and short loops, and 14 long loops exposed on the exterior cell surface [109]. Based on this model, MtrB forms a pore of 30–40 Å in diameter that can easily fit MtrA [32, 109]. The N-terminus of MtrB from *S. oneidensis* MR-1 and metal reducing *Shewanella* and *Ferrimonas* species contains a conserved CXXC motif that appears to be important for metal reduction [108]. Substitution of the first cysteine in the *S. oneidensis* MR-1 CXXC motif, C42, with an alanine, results in complete loss of metal reduction [108]. Substitution of both cysteine residues in this motif with serines also led to complete loss of metal reduction, and the mutagenized MtrB was not detected in Western blots, likely due to degradation. These results suggest that the N-terminus CXXC motif is important for stability of MtrB (Saffarini and Beliaev, unpublished results).

2.3.4 MtrC: An Outer Membrane Decaheme c-type Cytochrome

MtrC was first identified in 2001 as a major contributor to metal reduction [3] and it is the most studied component of the metal reductase. It is a decaheme *c*-type cytochrome located on the outer surface of *S. oneidensis* MR-1 cells and it transfers electrons directly to metal oxides and electrodes of microbial fuel cells [11, 12, 91, 110]. The external location of MtrC on the cell surface was determined by proteinase K treatment of whole cells or MtrC-containing liposomes and by atomic force microscopy [10, 56, 109]. MtrC is a lipoprotein [71] with a conserved N-terminal sequence (CGGS) found in MtrC proteins from other *Shewanella* species. The cysteine acts as the lipid attachment site and its replacement with a serine leads to accumulation of soluble MtrC in the culture supernatant [100] (Shirodkar and Saffarini, unpublished). Targeting MtrC to the outer cell surface requires the Type II secretion system, and mutants deficient in this system completely lose the ability to reduce metals [10]. MtrC is predicted to be a monomeric protein [42] with a uniform distribution on the surface of *S. oneidensis* MR-1 cells [56]. It also appears to have a slow turnover rate and is relatively insensitive to oxygen [111]. Biochemical analyses of MtrC indicate that it binds 10 low-spin hemes that are reduced within a potential window of +100 to −400 mV [42]. The crystal structure of MtrC has not been resolved, but a model was generated based on the structure of MtrF, an outer membrane decaheme *c*-type cytochrome described in more detail below. Based on this model, MtrC is predicted to have two domains with the hemes arranged in a staggered cross motif and in close proximity to each other thus facilitating electron transfer [27].

2.3.5 The Outer Membrane Cytochromes OmcA and MtrF

In addition to MtrC, the *S. oneidensis* MR-1 genome encodes two decaheme *c*-type cytochromes designated OmcA and MtrF. These proteins exhibit similarity to MtrC and their genes are located upstream of the *mtrCAB* operon. The participation of MtrF and OmcA in metal reduction has been extensively investigated and the results indicate that although both proteins are capable of metal reduction, their contribution to respiratory growth with Fe(III) appears to be minor. Mutants that lack MtrC exhibit a 75 % decrease in Fe(III) reduction compared to the wild type, suggesting that the residual activity is due to other outer membrane cytochromes. The contribution of OmcA and MtrF to metal reduction was determined using mutants that lack these genes. Deletion of *mtrF* or *omcA* did not result in a significant reduction in the mutants' ability to use metal oxides as electron acceptors. The role of MtrF and OmcA, however, became more evident in mutants that also lacked *mtrC*. Double (Δ*mtrC*Δ*omcA*) and triple (Δ*mtrC*Δ*omcA*Δ*mtrF*) were completely deficient in metal reduction [20] indicating that the residual metal reductase activity observed in *mtrC* mutants is due to the activity of OmcA, MtrF, or both. Interestingly, deletion of either *omcA* or *omcA* and *mtrF* in a Δ*mtrC* background gave comparable results with regard to metal reduction [20]. If MtrF contributes to metal reduction in vivo, one would expect the triple mutant to be more deficient in metal reduction than the Δ*mtrC*Δ*omcA* mutant. These results suggest that OmcA plays a bigger role than MtrF in metal reduction, perhaps accounting for the majority of the residual reductase activity observed in the Δ*mtrC* mutants. To further investigate the roles of OmcA and MtrF in metal reduction, mutants that lack all outer membrane *c*-type cytochromes were generated and transformed with medium to high copy-number plasmids carrying either *omcA* or *mtrF*. Introduction of *omcA* into these mutants did not restore metal reduction, in contrast to complementation with *mtrF* that allowed the mutant to reduce iron oxides to wild type levels [15, 20]. These results are surprising because MtrF is not known to play a significant role in metal reduction compared to OmcA. The inability of OmcA to restore metal reduction in the mutant was attributed to the absence of MtrC that is thought to transfer electrons to OmcA [15].

Purified OmcA has one high-spin and 9 low-spin hemes and can strongly bind to and reduce hematite and ferrihydrite [7, 46, 55, 64, 112]. It attaches to hematite in a confirmation that allows direct electron transfer through maximum contact with the metal [46]. OmcA also strongly interacts with MtrC to form a tight complex, and this interaction is thought to enhance metal reduction [64, 91, 99]. MtrF is predicted be a component of the MtrDEF complex that is similar to MtrCAB, but is postulated to have a function distinct from other outer membrane *c*-type cytochromes [82]. The *mtrDEF* genes are highly expressed in cell aggregates in response to calcium and it is suggested that MtrDEF play a role in detoxification or reduction of radionuclides under aerobic conditions [62]. The structures of OmcA and MtrF were recently determined at 2.7 and 3.2 Å respectively [18, 29]. Although these proteins share low sequence identity (25 %), their basic structure and heme arrangement appear to be similar. In both proteins, hemes are arranged in two chains that intersect and form a staggered cross motif. Each heme is within 7 Å from its nearest neighbor thus allowing rapid electron transfer between hemes [18, 29].

Although the genomes of several *Shewanella* species contain *omcA* and *mtrF*, others lack these genes despite the fact that these species are able to reduce metals similar to *S. oneidensis* MR-1 [36]. Examples include *S. putrefaciens* W-3-18-1 and *Shewanella* sp. strain HRCR-6 that express, in addition to MtrC, an outer membrane *c*-type cytochrome designated UndA. This protein is an 11-heme *c*-type cytochrome that appears to play a role in metal reduction, can partially restore Fe (III) reductase activity to an *S. oneidensis* MR-1 Δ*mtrC*Δ*omcA* mutant, and appears to have uranium (VI) reductase activity [98, 115]. The crystal structure of UndA from strain HRCR-6 was recently determined at 1.67 Å [28]. Despite the differences in the number of heme *c* groups that each protein binds, a comparison of UndA and MtrF structures indicated they share a conserved 10 heme staggered cross motif [28], similar to OmcA and possibly MtrC.

2.4 Electron Shuttles and Microbial Nanowires

There has been much debate as to whether *S. oneidensis* MR-1 reduces metals directly through contact of the reductase with the metal oxides or through intermediates such as electron shuttles and nanowires. Electron shuttles are soluble redox-active molecules that can mediate electron transfer between the cell surface and metal oxides or electrodes. The role of electron shuttles in metal reduction gained interest following the report by Newman and Kolter [77] who documented the potential involvement of excreted quinones in extracellular electron transfer [77]. In 2008, two groups identified riboflavin and flavin mononucleotide (FMN) as the electron shuttling molecules in *S. oneidensis* MR-1 and other *Shewanella* species [61, 107]. Although flavin secretion is thought to be important for metal reduction, which occurs under anaerobic conditions, flavin concentrations were similar in supernatants of aerobic and anaerobic cultures [107]. All 23 sequenced *Shewanella* species have the genes for riboflavin biosynthesis, including *S. denitrificans* that secretes flavins but does not reduce metals [14]. Flavins have been shown to accelerate electron transfer to metals oxides and are thought to allow greater access to these electron acceptors [1, 18, 50, 61, 107]. Kotloski and Gralnick isolated a mutant that lacks the bacterial FAD exporter (Bfe; SO_0702) and determined that this protein is involved in flavin secretion [50]. The *bfe* mutant was severely impaired in ferrihydrite reduction but was able to reduce ferric citrate similar to the wild type. Based on their results, the authors suggested that flavins account for 75 % of insoluble metal reduction activity under their laboratory conditions, while the rest is due to direct contact of cells with metal oxides [50]. The slower rate of metal reduction in the absence of flavins led to the hypothesis that the activity of MtrC and OmcA results in a bottleneck in the electron transfer pathway that is relieved by redox active molecules such as flavins [1, 90]. Recently however, Okamoto and colleagues demonstrated that acceleration of electron transfer by free flavins is not energetically favorable [78]. Rather, flavins associate directly with outer membrane *c* cytochromes as semiquinone cofactors that mediate one-electron transfer reactions [78, 79]. A similar mechanism appears to operate in the anaerobic metal reducer *Geobacter sulfurreducens* [80, 81]. Flavin binding to MtrC and OmcA exhibits specificity where FMN binds to MtrC and riboflavin associates with OmcA [78, 79].

In addition to flavins, metal reducing bacteria are thought to use appendages, called nanowires, to transfer electrons to metal oxides and electrodes of microbial fuel cells at a distance. In *S. oneidensis* MR-1 cells grown under limited oxygen conditions, electrically conductive appendages, or nanowires, were detected using scanning tunneling microscopy and tunneling spectroscopy [39]. Mutants that lack outer membrane *c*-type cytochromes, and therefore are deficient in metal reduction, produced non-conductive nanowires [30, 39]. Electrically conductive nanowires were also identified in the metal reducer *G. sulfurreducens* [86]. In this bacterium, PilA, the major component of type IV pili, was found to be electrically conductive, and its loss resulted in loss of iron oxide reduction [86]. Aromatic amino acids in the *G. sulfurreducens* PilA C-terminus appear to be important for PilA function, and their substitution with other amino acids decreases conductivity and metal reduction [105]. *S. oneidensis* MR-1 expresses two types of pili, Type IV and Msh, on its cell surface [11]. To determine the role of these appendages in metal reduction and electricity production, we generated mutants that are deficient in type IV and Msh pili biogenesis (Fig. 2.1a). The mutants were able to reduce metal oxides similar to the wild type (Fig. 2.1b) and generate electricity in microbial fuel cells [11], suggesting that unlike *G. sulfurreducens* pili, the *S. oneidensis* MR-1 pili are not involved in long-range extracellular electron transfer. Recently, the nature of the conductive appendages or nanowires in *S. oneidensis* MR-1 was revealed [83]. Pirbadian and colleagues were able to show in real time the formation of *S. oneidensis* MR-1 nanowires. Using immuno-fluorescence imaging, the authors were able to monitor the formation of the nanowires and show that they are extensions, or vesicles, of the outer membrane and periplasm [83]. These vesicles would be expected, therefore, to contain outer membrane *c*-type cytochromes that are capable of extracellular electron transfer.

2.5 Extracellular Electron Transfer by *S. oneidensis* MR-1

The localization of the metal reductase in the outer membrane of *S. oneidensis* MR-1 cells with exposure to the extracellular environment requires electron transfer from the inner membrane, through the periplasmic space and outer membrane, and finally to the extracellular electron acceptor. Following the identification of the MtrCAB proteins, a simple model that describes this electron transport chain was proposed [3]. Since then, a wealth of data elucidated protein-protein interactions, protein localization and structure, and provided a more detailed and refined model of the mechanism by which *S. ondeidensis* MR-1 transfers electrons extracellularly to metal oxides. Five proteins have been confirmed through biochemical and genetic analyses to be involved in metal reduction. These consist of MtrB, MtrC, OmcA, MtrA, and CymA (Fig. 2.2). The core metal reductase components consist of MtrCAB. These proteins confer metal reductase activity on *E. coli* [45] and are sufficient to account for physiological levels of metal reduction in *S. oneidensis* MR-1 [109]. MtrA is a periplasmic protein that forms tight interactions with MtrB and MtrC [91] and is embedded within MtrB forming a "porin cytochrome" electron transfer module [88]. MtrC is exposed on the outer cell surface, and presumably interacts with MtrA within the MtrB pore. Our understanding of the interactions between MtrA and MtrC within the porin cytochrome model is sufficient to explain electron transfer from MtrA to MtrC and subsequently to extracellular electron acceptors.

OmcA, similar to MtrC, is a decaheme *c*-type cytochrome that is exposed on the cell surface and requires the type II secretion system to reach its final destination. Analysis of OmcA crystals suggests that the protein forms a dimer in the outer membrane [29], and cross-linking experiments identified an MtrC/OmcA complex

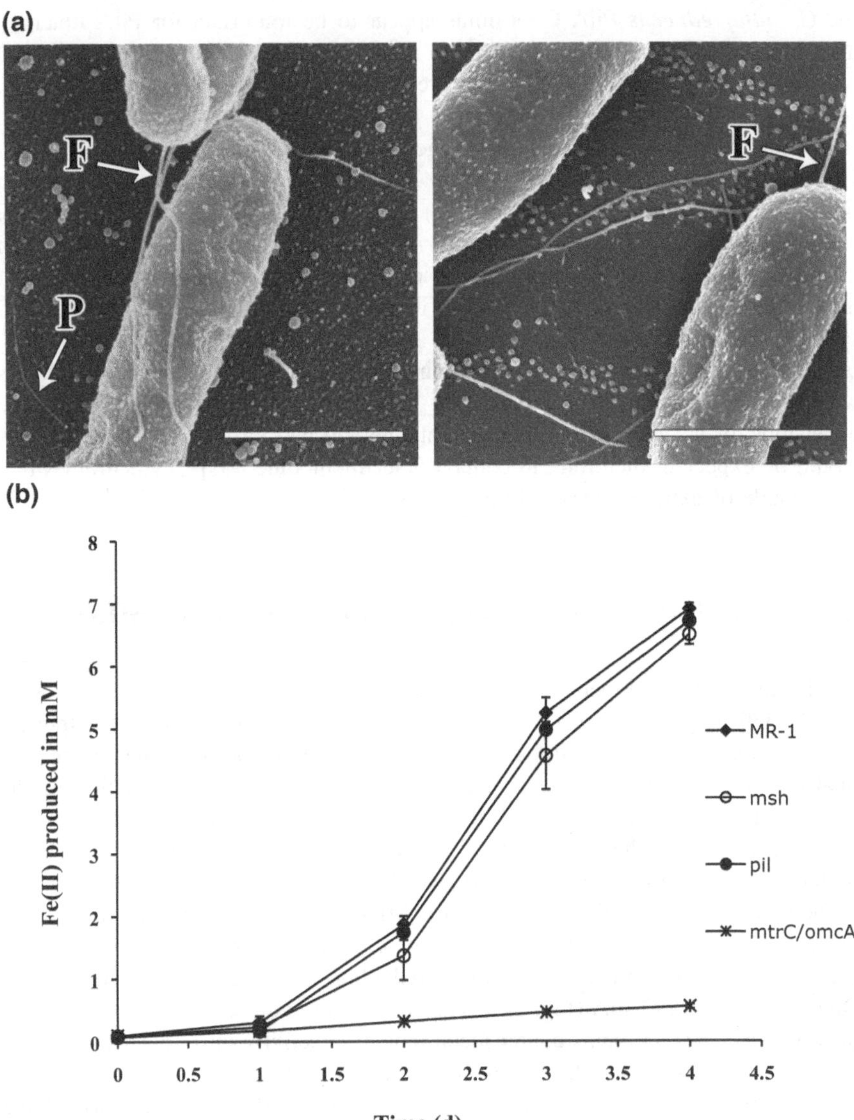

Fig. 2.1 Role of *S. oneidensis* MR-1 pili in metal reduction. **a** Scanning electron micrograph of *S. oneidensis* MR-1 (*left panel*) and Δ*pil*Δ*msh* mutant that lacks the type IV and Msh pili biogenesis systems (*right panel*). Flagella and pili are indicated. *White bar* = 1 μm. **b** Iron oxide reduction by *S. oneidensis* MR-1 and mutants strains. In contrast to the mutant that lacks MtrC and OmcA, mutants deficient in type IV or Msh pili biogenesis (Δ*pil* and Δ*msh*) reduced iron oxide similar to the wild type (see [11] for more details)

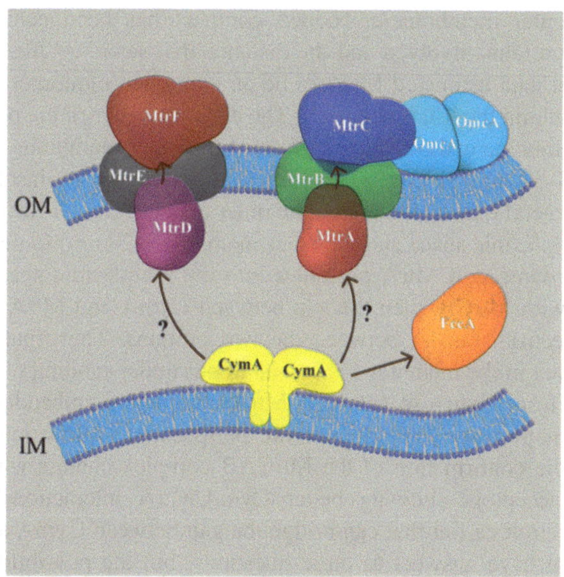

Fig. 2.2 Model of the *S. oneidensis* MR-1 electron transport chain that leads to extracellular metal reduction. CymA is predicted to be a dimer and a quinol oxidoreductase that transfers electrons from the inner membrane to MtrA. It is not clear at present if an intermediate electron carrier links CymA with the decaheme periplasmic proteins MtrA. The core metal reductase complex consists of MtrABC, where MtrC is exposed on the cell surface and MtrA is embedded in the MtrB pore and transfers electrons from the periplasm to MtrC. OmcA participates in metal reduction and is predicted to be a dimer. MtrDEF is a second outer membrane complex that is similar to MtrCAB, but its function in metal reduction is not clear (see text for more detail)

in a 2:1 ratio [29, 112, 117] in support of the oligomeric state of OmcA. Although OmcA reduces metal oxides, this reduction does not appear to contribute to bacterial growth in contrast to metal reduction by MtrC [65]. OmcA is thought to receive its electrons from reduced MtrC [99], yet *mtrC* mutants can still carry out metal reduction. This discrepancy can be explained by two possibilities. OmcA may be reduced by MtrA and not MtrC in vivo, or a yet to be identified electron carrier is responsible for reducing OmcA in the absence of MtrC. Although the mechanisms that lead to OmcA reduction warrant further investigation, it is clear that MtrC and OmcA participate in extracellular electron transfer to metal oxides and electrodes of microbial fuel cells.

CymA, a membrane anchored *c*-type cytochrome, is the only confirmed link to date between the menaquinol pool and the metal reductase MtrCAB. CymA, as mentioned above, is a menaquinol oxidase predicted to form a homodimer [9, 58] and is anchored to the inner membrane facing the periplasm (Fig. 2.2). Use of a bacterial two-hybrid system provided evidence for the interaction of CymA with MtrA [9]. This interaction appears to be transient and leads to MtrA reduction [33]. Direct electron transfer, however, from CymA in the inner membrane to MtrA that is part of an outer membrane-embedded complex has been debated given the dimensions of proteins involved and the distance that separates them. Small-angle X-ray scattering data estimated MtrA to be an elongated monomer of 10.4 nm in length that fits within the MtrB pore [32]. The distance between the periplasmic side of the inner membrane and the outer surface of the outer membrane is estimated to be roughly 28–30 nm [24, 101]. Given that reduction of insoluble electron acceptors (i.e., metal oxides) occurs on the outer cell surface, electron carriers must traverse the periplasmic space and the outer membrane (28 nm) to deliver electrons to MtrC and/or OmcA. If MtrA protrudes into the periplasmic space but forms a tight complex with MtrCB, then the gap between CymA and MtrA is too wide to allow direct electron transfer between the two proteins. The organization of the MtrCAB complex in the membrane is crucial to our understanding of how electrons are transferred from CymA to MtrA. Is MtrA completely embedded in the MtrB pore, or does it protrude enough into the periplasm to allow interactions with CymA? Does the confirmation of the MtrCAB complex change when it interacts with electron acceptors allowing better CymA/MtrA interactions? Is there an unidentified electron carrier that can bridge the gap between CymA and MtrA? We currently do not have answers to these questions, but the possibility of an intermediate electron carrier exists and two *c*-type cytochromes, FccA and STC, have been proposed to serve as intermediates that transfer electron from CymA to MtrC. Strong evidence, however, to support the involvement of these proteins in metal reduction is lacking. FccA is a flavocytochrome *c* with confirmed fumarate reductase activity [51, 57, 92], whereas STC is a small tetraheme *c*-type cytochrome that appears to bind chelated Fe(III) in vitro but its function in vivo has not been determined [85, 104]. Experiments with bacterial two-hybrid systems suggested that FccA and STC interact with MtrA, but interactions between STC and CymA were not detected [9]. In contrast, using NMR spectroscopy, Fonseca et al. [34] suggested that FccA and STC transiently interact with CymA and MtrA thus acting as the bridge in the periplasmic electron transfer to the OM. This finding would predict that mutants deficient in STC or FccA are impaired in metal reduction. In contrast to this notion, deletion of *cctA* that encodes STC does not affect the ability of *S. oneidensis* MR-1 cells to reduce metals [12]. Furthermore, deletion of *fccA* leads to increased metal reductase activity [95]. These findings led Fonseca et al. [34] to predict that STC and FccA have redundant functions. In the absence of double mutants that lack metal reductase activity, we cannot conclude that either STC or FccA participate in electron transfer to MtrA.

2.6 Concluding Remarks

Since the isolation of *S. oneidensis* and *G. metallireducens* in pure culture in 1988, intensive investigation and a wealth of data provided in depth insight into the physiology, biochemistry, and genetics of dissimilatory metal reducing bacteria. *S. oneidensis* is a model organism for studying metal reduction, and in this chapter we focused mainly on the key components used by this organism to enable extracellular respiration. Unlike other respiratory pathways, the metal reducing electron transport chain extends from the inner membrane to the outer cell surface in a process that has become known as extracellular electron transfer. Several mechanisms have been proposed to explain this process in *S. oneidensis* and other bacteria. These include (i) conductive nanowires, (ii) production of soluble extracellular electron shuttles, and (iii) direct contact of bacterial cells with the insoluble metals. Conductive extracellular appendages, or nanowires, have been shown recently to be outer membrane vesicles that would contain metal reductase components. The commonly used term of nanowires to describe these vesicles does not accurately reflect the nature of these structures. Once these vesicles separate from the cell, and without a continuous source of electrons from the cytoplasm, they will be incapable of extracellular electron transfer. Flavin electron shuttles, that were thought to allow access to metal oxides at a distance, function as cofactors that bind to the outer membrane cytochromes and mediate electron transfer. Thus it appears that *S. oneidensis* reduces metal oxides and electrodes of microbial fuel cells mostly by direct contact. In spite of several major breakthroughs described in this review, gaps in our understanding of the metal reduction pathway still exist. The precise molecular structures of the periplasmic and outer membrane electron transport complexes as well as the biogenesis of the metal reductase complex will be crucial to further our understanding of extracellular electron transport. It is noteworthy that extracellular electron transfer is not unique to *S. oneidensis* and is prevalent in bacterial and archaeal species. Understanding the mechanisms that underlie this process not only sheds light on an unusual yet widespread and environmentally significant microbial activity, it also allows us to better design and use these microorganisms in a variety of applications that range from bioremediation of contaminated subsurface environments to electricity and biofuel production.

References

1. Baron D, LaBelle E, Coursolle D, Gralnick J, Bond D (2009) Electrochemical measurement of electron transfer kinetics by *Shewanella oneidensis* MR-1. J Biol Chem 284:28865–28873
2. Beliaev A, Saffarini D (1998) *Shewanella putrefaciens mtrB* encodes an outer membrane protein required for Fe(III) and Mn(IV) reduction. J Bacteriol 180:6292–6297
3. Beliaev A, Saffarini D, McLaughlin J, Hunnicut D (2001) MtrC, an outer membrane decaheme *c* cytochrome required for metal reduction in *Shewanella putrefaciens* MR-1. Mol Microbiol 39:722–730
4. Beliaev A, Thompson D, Khare T, Lim H, Brandt C, Li G, Murray A, Heidelberg J, Giometti C, Yates J, Nealson K, Tiedje J, Zhou G (2002) Gene and protein expression profiles of *Shewanella oneidensis* during anaerobic growth with different electron acceptors. OMICS 6
5. Beliaev AS, Klingeman DM, Klappenbach JA, Wu L, Romine MF, Tiedje JM, Nealson KH, Fredrickson JK, Zhou J (2005) Global transcriptome analysis of *Shewanella oneidensis* MR-1 exposed to different terminal electron acceptors. J Bacteriol 187:7138–7145

6. Bencheikh-Latmani R, Williams SM, Haucke L, Criddle CS, Wu L, Zhou J, Tebo BM (2005) Global transcriptional profiling of *Shewanella oneidensis* MR-1 during Cr(VI) and U(VI) reduction. Appl Environ Microbiol 71:7453–7460
7. Bodemer G, Antholine W, Basova L, Saffarini D, Pacheco I (2010) The effect of detergents and lipids on the properties of the outer-membrane OmcA from *Shewanella oneidensis*. J Biol Inorg Chem 15:749–758
8. Bordi C, Ansaldi M, Gon S, Jourlin-Castelli C, Iobbi-Nivol C, Mejean V (2004) Genes regulated by TorR, the trimethylamine oxide response regulator of *Shewanella oneidensis*. J Bacteriol 186:4502–4509
9. Borloo J, de Smet L, Van Beeumen J, Devreese B (2011) Bacterial two-hybrid analysis of the *Shewanella oneidensis* MR-1 multi-component electron transfer pathway JIOMICS 1:260–267
10. Bouhenni R (2007) Investigation of the mechanisms of iron(III) and manganese(IV) reduction in *Shewanella oneidensis* MR-1. University of Wisconsin-Milwaukee, Milwaukee
11. Bouhenni R, Vora G, Biffinger J, Shirodkar S, Brockman K, Ray R, Wu P, Johnson B, Biddle E, Marshall M, Fizgerald L, Little B, Fredrickson J, Beliaev A, Ringeison B, Saffarini D (2010) The role of *Shewanella oneidesis* outer surface structures in extracellular electron transfer. Electroanalysis 22:856–864
12. Bretschger O, Obraztsova A, Sturm CA, Chang IS, Gorby YA, Reed SB, Culley DE, Reardon CL, Barua S, Romine MF, Zhou J, Beliaev AS, Bouhenni R, Saffarini D, Mansfeld F, Kim BH, Fredrickson JK, Nealson KH (2007) Current production and metal oxide reduction by *Shewanella oneidensis* MR-1 wild type and mutants. Appl Environ Microbiol 73:7003–7012
13. Brown R, Romine M, Schepmoes A, Smith RD, Lipton MS (2010) Mapping the subcellular proteome of *Shewanella oneidensis* MR-1 using sarkosyl-based fractionation and LC-MS/MS identification. J Proteome Res 9:4454–4463
14. Brutinel E, Gralnick J (2012) Shuttling happens: soluble flavin mediators of extracellular electron transfer in *Shewanella*. Appl Microbiol Biotechnol 93:41–48
15. Bucking C, Popp F, Kerzenmacher S, Gescher JS (2010) Involvement and specificity of *Shewanella oneidensis* outer membrane cytochromes in the reduction of soluble and solid-phase terminal electron acceptors. FEMS Micorbiol Lett 306:144–151
16. Burns J, DiChristina T (2009) Anaerobic respiration of elemental sulfur and thiosulfate bu *Shewanella oneidensis* MR-1 requires *psrA*, a homolog of the *phs* gene of *Salmonella enterica* serovar *typhimurium* LT2. Appl Environ Microbiol 75:5209–5217
17. Charania M, Brockman K, Zhang Y, Banerjee A, Pinchuk G, Fredrickson J, Beliaev A, Saffarini D (2009) Involvement of a membrane-bound class III adenylate cyclase in the regulation of anaerobic respiration in *Shewanella oneidensis* MR-1. J Bacteriol 191:4298–4306
18. Clarke T, Edwards M, Gates A, Hall A, White G, Bradley J, Reardon CL, Shi L, Beliaev A, Marshall M, Wang Z, Watmough N, Fredrickson J, Zachara J, Butt J, Richardson D (2011) Structure of a bacterial cell surface decaheme electron conduit. Proc Natl Acad Sci USA 108:9384–9389
19. Cordova CD, Schicklberger M, Yu Y, Spormann A (2011) Partial functional replacement of CymA by SirCD in *Shewanella oneidensis* MR-1. J Bacteriol 193:2312–2321
20. Coursolle D, Gralnick J (2012) Reconstruction of extracellular respiratory pathways for iron (III) reduction in *Shewanella oneidensis* strain MR-1. Front Microbiol 3:11
21. Cruz-Garcia C, Murray AE, Klappenbach JA, Stewart V, Tiedje J (2007) Respiratory nitrate ammonification by *Shewanella oneidensis* MR-1. J Bacteriol 189:656–662
22. Cummings D, Caccavo F, Spring S, Rosenzweig R (1999) *Ferribacterium limneticum*, gen. nov., sp. nov., an Fe(III)-reducing miroorganism isolated from mining-impacted freshwater lake sediment. Arch Microbiol 171:183–188
23. Cummings D, March A, Bostick B, Spring S, Caccavo F, Fendorf S, Rosenzweig R (2000) Evidence of microbial Fe(III) reduction in anoxic mining-impacted lake sediments (Lake Coeur d'Alene, Idaho). Appl Environ Microbiol 66:154–162

24. Dohnalkova A, Marshall M, Arey B, Williams K, Buck E, Fredrickson J (2011) Imaging hydrated microbial extracellular polymers: comparative analysis by electron microscopy. Appl Environ Microbiol 77:1254–1262
25. Dos Santos J, Iobbi-Nivol C, Couillault C, Giordano G, Mejean V (1998) Molecular analysis of the trimethylamine N-oxide (TMAO) reductase respiratory system from a *Shewanella species*. J Mol Biol 284:421–433
26. Duncan G, McMillan G, Marritt S, Firer-Sherwood M, Shi L, Richardson D, Evans S, Elliott S, Butt J, Jeuken L (2014) Protein-protein interaction regulates the direction of catalysis and electron transfer in a redox enzyme complex. J Am Chem Soc 135:10550–10556
27. Edwards M, Fredrickson J, Zachara J, Richardson D, Clarke T (2012) Analysis of structural MtrC models based on homology with the crystal structure of MtrF. Biochem Soc Trans 40:1181–1185
28. Edwards M, Hall A, Shi L, Fredrickson J, Zachara J, Butt J, Richardson D, Clarke T (2012) The crystal structure of the extracellular 11-heme cytochrome UndA reveals a conserved 10-heme motif and defined binding site for soluble iron chelates. Structure 20:1275–1284
29. Edwards M, Nanakow B, Johs A, Tomanicek S, Liang L, Shi L, Fredrickson J, Zachara J, Gates A, Butt J, Richardson D, Clark M (2014) The X-ray crystal structure of *Shewanella oneidensis* OmcA reveals new insight at the microbe-metal interface. FEBS Lett 588:1886–1890
30. El-Naggar M, Wagner G, Leung K, Yuzvinsky T, Southam G, Yang J, Lau W, Nealson K, Gorby U (2010) Electrical transport along bacterial nanowires from *Shewanella oneidensis* MR-1. Proc Natl Acad Sci USA 107:18127–18131
31. Emerson D, Roden E, Twining B (2012) The microbial ferrous wheel: iron cycling in terrestrial, freshwater, and marine environments. Front Microbiol 3:383
32. Firer-Sherwood M, Ando N, Drennan C, Elliott S (2011) Solution-based structural analysis of the decaheme cytochrome, MtrA, by small-angle-X-ray scattering and analytical ultracentrifugation. J Phys Chem 115:11208–11214
33. Firer-Sherwood M, Bewley K, Mock J-Y, Elliott S (2011) Tools for resolving complexity in the electron transfer netwroks of multiheme cytochromes *c*. Metallomics 3:344–348
34. Fonseca B, Paquete C, Neto S, Pacheco I, Soares C, Louro R (2013) Mind the gap: cytochrome interactions reveal electron pathways across the periplasm of *Shewanella oneidensis* MR-1. Biochem J 449:101–108
35. Francis CA, Obraztsova AY, Tebo BM (2000) Dissimilatory metal reduction by the facultative anaerobe *Pantoea agglomerans* SP1. Appl Environ Microbiol 66:543–548
36. Fredrickson JK, Romine MF, Beliaev AS, Auchtung JM, Driscoll ME, Gardner TS, Nealson KH, Osterman AL, Pinchuk G, Reed JL, Rodionov DA, Rodrigues JL, Saffarini DA, Serres MH, Spormann AM, Zhulin IB, Tiedje JM (2008) Towards environmental systems biology of *Shewanella*. Nat Rev Microbiol 6:592–603
37. Gao H, Yang Z, Barua S, Reed S, Romine M, Nealson K, Fredrickson J, Tiedje J, Zhou J (2009) Reduction of nitrate in *Shewanella oneidensis* depends on atypical NAP and NRF systems with NapB as a preferred electron transport protein from CymA to NapA. ISME J 3:966–976
38. Gon S, Patte J, Dos Santos J, Mejean V (2002) Reconstitution of the trimethylamine oxide reductase regulatory elements of *Shewanella oneidensis* in *Escherichia coli*. J Bacteriol 184:1262–1269
39. Gorby YA, Yanina S, McLean JS, Rosso KM, Moyles D, Dohnalkova A, Beveridge TJ, Chang IS, Kim BH, Kim KS, Culley DE, Reed SB, Romine MF, Saffarini DA, Hill EA, Shi L, Elias DA, Kennedy DW, Pinchuk G, Watanabe K, Ishii S, Logan B, Nealson KH, Fredrickson JK (2006) Electrically conductive bacterial nanowires produced by *Shewanella oneidensis* strain MR-1 and other microorganisms. Proc Natl Acad Sci USA 103:11358–11363
40. Gralnick JA, Newman D (2007) Extracellular respiration. Mol Microbiol 65:1–11
41. Gralnick JA, Vali H, Lies DP, Newman DK (2006) Extracellular respiration of dimethyl sulfoxide by *Shewanella oneidensis* strain MR-1. Proc Natl Acad Sci USA 103:4669–4674

42. Hartshorne RS, Jepson BN, Clarke TA, Field SJ, Fredrickson J, Zachara J, Shi L, Butt JN, Richardson DJ (2007) Characterization of *Shewanella oneidensis* MtrC: a cell-surface decaheme cytochrome involved in respiratory electron transport to extracellular electron acceptors. J Biol Inorg Chem 12:1083–1094
43. Hau H, Gralnick J (2007) Ecology and biotechnology of the genus *Shewanella*. Annu Rev Microbiol 61:237–258
44. Heidelberg J, Paulsen I, Nealson K, Gaidos E, Nelson W, Read T et al (2002) Genome sequence of the dissimilatory metal ion-reducing bacterium *Shewanella oneidensis*. Nat Biotechnol 20:1118–1123
45. Jensen H, Albers A, Malley K, Londer Y, Cohen B, Helms B, Weigele P, Grove J, Ajo-Franklin C (2010) Engineering of a synthetic electron conduit in living cells. Proc Natl Acad Sci USA 107:19213–19218
46. Johs A, Droubay T, Ankner J, Liang L (2010) Characterization of the decaheme *c*-type cytochrome OmcA in solution and on hematite surfaces by small angle X-ray scattering and neutron refratometry. Biophys J 98:3035–3043
47. Kashefi K, Tor JM, Holmes DE, Gaw Van Praagh CV, Reysenbach AL, Lovley DR (2002) *Geoglobus ahangari* gen. nov., sp. nov., a novel hyperthermophilic archaeon capable of oxidizing organic acids and growing autotrophically on hydrogen with Fe(III) serving as the sole electron acceptor. Int J Syst Evol Microbiol 52:719–728
48. Kieft TL, Fredrickson JK, Onstott TC, Gorby YA, Kostandarithes HM, Bailey TJ, Kennedy DW, Li SW, Plymale AE, Spadoni CM, Gray MS (1999) Dissimilatory reduction of Fe(III) and other electron acceptors by a *Thermus* isolate. Appl Environ Microbiol 65:1214–1221
49. Kirchman D (1996) Microbial Ferrous Wheel. Nature 383:303–304
50. Kotloski N, Gralnick J (2013) Flavin electron shuttles dominate extracellular electron transfer by *Shewanella oneidensis*. mBio 4:e00553–e00612
51. Leys D, Tsapin A, Nealson K, Meyer T, Cusanovich M, Van Beeumen J (1999) Structure and mechanisms of the flavocytochrome *c* reductase of *Shewanella putrefaciens* MR-1. Nat Struct Biol 6:1113–1117
52. Liu C, Gorby YA, Zachara JM, Fredrickson JK, Brown CF (2002) Reduction kinetics of Fe (III), Co(III), U(VI), Cr(VI), and Tc(VII) in cultures of dissimilatory metal-reducing bacteria. Biotechnol Bioeng 80:637–649
53. Lovley D (2006) Dissimilatory Fe(III)- and Mn(IV)-reducing prokaryotes. In Dworkin M, Falkow S, Rosenberg E, Schleifer K, Stackebrandt E (eds) The prokaryotes, vol. 2. Springer, Berlin, pp 635–658
54. Lovley D, Phillips E (1988) Novel mode of microbial energy metabolism: organic carbon oxidation coupled to dissimilatory reduction of iron and manganese. Appl Environ Microbiol 54:1472–1480
55. Lower B, Shi L, Yongsunthon R, Droubay T, McCready D, Lower SK (2007) Specific bonds between an iron oxide surface and outer membrane cytochromes MtrC and OmcA from *Shewanella oneidensis* MR-1. J Bacteriol 189:4944–4952
56. Lower B, Yongsunthon R, Shi L, Wilding L, Gruber H, Wigginton NS, Reardon CL, Pinchuk G, Droubay T, Boily J, Lower SK (2009) Antibody recognition force microscopy shows that outer membrane cytochromes OmcA and MtrC are expressed on the exterior surface of *Shewanella oneidensis* MR-1. Appl Environ Microbiol 75:2931–2935
57. Maier T, Myers J, Myers C (2003) Identification of the gene encoding the sole physiological fumarate reductase in *Shewanella oneidensis* MR-1. J Basic Micorbiol 43:312–327
58. Marritt S, Lowe T, Bye J, McMillan D, Shi L, Fredrickson J, Zachara J, Richardson D, Cheesman M, Jeuken L, Butt J (2012) A functional description of CymA, an electron-transfer hub supporting anaerobic respiratory flexibility in *Shewanella*. Biochem J 444:465–474
59. Marshall MJ, Beliaev AS, Dohnalkova AC, Kennedy DW, Shi L, Wang Z, Boyanov MI, Lai B, Kemner KM, McLean JS, Reed SB, Culley DE, Bailey VL, Simonson CJ, Saffarini DA, Romine MF, Zachara JM, Fredrickson JK (2006) *c*-Type cytochrome-dependent formation of U(IV) nanoparticles by *Shewanella oneidensis*. PLoS Biol 4:e268

60. Marshall MJ, Plymale AE, Kennedy DW, Shi L, Wang Z, Reed SB, Dohnalkova AC, Simonson CJ, Liu C, Saffarini DA, Romine MF, Zachara JM, Beliaev AS, Fredrickson JK (2008) Hydrogenase- and outer membrane c-type cytochrome-facilitated reduction of technetium(VII) by *Shewanella oneidensis* MR-1. Environ Microbiol 10:125–136
61. Marsili E, Baron D, Shikhare I, Coursolle D, Gralnick J, Bond D (2008) *Shewanella* secretes flavins that mediate extracellular electron transfer. Proc Natl Acad Sci USA 105:3968–3973
62. McLean J, Pinchuk G, Geydebrekht O, Bilskis C, Zakrajsek B, Hill E, Saffarini D, Romine M, Gorby Y, Fredrickson J, Beliaev A (2008) Oxygen-dependent autoaggregation in *Shewanella oneidensis* MR-1. Environ Microbiol 10:1861–1876
63. McMillan D, Marritt S, Butt J, Jeuken L (2012) Menaquinone-7 is specific cofactor in tetraheme quinol dehydrogenase CymA. J Biol Chem 287:14215–14225
64. Meitl L, Eggleston C, Colberg P, Khare N, Reardon CL, Shi L (2009) Electrochemical interaction of *Shewanella oneidensis* MR-1 and its outer membrane cytochrome OmcA and MtrC with hematite electrodes. Geochim Cosmochim Acta 73:5292–5307
65. Mitchell A, Peterson L, Reardon C, Reed S, Culley D, Romine M, Geesey G (2012) Role of outer membrane *c*-type cytochromes MtrC and OmcA in *Shewanella oneidensis* MR-1 cell production, accumulation, and detachment during respiration on hematite. Geobiol 10:355–370
66. Morris CJ, Black A, Pealing S, Manson F, Chapman SK, Reid G, Gibson D, Ward FB (1944) Purification and properties of a novel cytochrome: flavocytochrome *c* from *Shewanella putrefaciens*. Biochem J 302:587–593
67. Moser D, Nealson K (1996) Growth of the facultative anaerobe *Shewanella putrefaciens* by elemental sulfur reduction. Appl Environ Microbiol 62:2100–2105
68. Murphy J, Durbin K, Saltikov C (2009) The functional roles of *arcA*, *etrA*, cyclic AMP (cAMP)-cAMP receptor protein, and *cya* in the arsenate respiration pathway in *Shewanella* sp. strain ANA-3. J Bacteriol 191:1035–1043
69. Murphy JN, Saltikov CW (2007) The cymA gene, encoding a tetraheme c-type cytochrome, is required for arsenate respiration in *Shewanella* species. J Bacteriol 189:2283–2290
70. Myers C, Nealson K (1988) Bacterial manganese reduction and growth with manganese oxide as the sole electron acceptor. Science 240:1319–1321
71. Myers CR, Myers JM (2003) Cell surface exposure of the outer membrane cytochromes of *Shewanella oneidensis* MR-1. Lett Appl Microbiol 37:254–258
72. Myers CR, Myers JM (1997) Cloning and sequence of *cymA*, a gene encoding tetraheme cytochrome *c* required for reduction of iron(III), fumarate, and nitrate by *Shewanella putrefaciens* MR-1. J Bacteriol 179:1143–1152
73. Myers J, Myers C (2000) Chromium (VI) reductase activity is associated with the cytoplasmic membrane of anaerobically grown *Shewanella putrefaciens* MR-1. J Appl Microbiol 88:98–106
74. Myers J, Myers C (2000) Role of the tetraheme cytochrome CymA in anaerobic electron transport in cells of *Shewanella putrefaciens* MR-1 with normal levels of menaquinone. J Bacteriol 182:67–75
75. Myers JM, Antholine WE, Myers CR (2004) Vanadium(V) reduction by *Shewanella oneidensis* MR-1 requires menaquinone and cytochromes from the cytoplasmic and outer membranes. Appl Environ Microbiol 70:1405–1412
76. Nealson K, Saffarini D (1994) Iron and Manganese in anaerobic respiration: environmental significance, physiology, and regulation. Ann Rev Microbiol 48:311–343
77. Newman DK, Kolter R (2000) A role for excreted quinones in extracellular electron transfer. Nature 405:94–97
78. Okamoto A, Hashimoto K, Nealson K, Nakamura R (2013) Rate enhancement of bacterial extracellular electron transport involved bound flavin semiquinones. Proc Natl Acad Sci USA 110:7856–7861
79. Okamoto A, Kalathil S, Deng X, Hashimoto K, Nakamura R, Nealson K (2014) Cell-secreted flavins bound to membrane cytochromes dictate electron transfer reactions to surfaces with diverse charge and pH. Sci Rep 4:5628

80. Okamoto A, Nakamura R, Nealson K, Hashimoto K (2014) Bound flavin model suggests similar electron-transfer mechanisms in *Shewanella* and *Geobacter*. ChemElectroChem 1:1808–1812
81. Okamoto A, Saito K, Inoue K, Nealson K, Hashimoto K, Nakamura R (2014) Uptake of self-secreted flavins as bound cofactors for extracellular electron transfer in *Geobacter* species. Energy Environ Sci 7:1357
82. Paquete C, Fonseca B, Cruz D, Periera T, Pacheco I, Soares C, Louro R (2014) Exploring the molecular mechanisms of electron shuttling across the microbe/metal space. Front Microbiol 5:318
83. Pirbadian S, Barchinger S, Leung K, Byun H, Jangir Y, Bouhenni R, Reed S, Romine M, Saffarini D, Shi L, Gorby U, Golbeck J, El-Naggar M (2014) *Shewanella oneidensis* MR-1 nanowires are outer membrane and periplasmic extensions of the extracellular electron transport components. Proc Natl Acad Sci USA 111:12883–12888
84. Pitts K, Dobbin P, Reyes-Ramirez F, Thomson A, Richardson D, Seward H (2003) Characterizaton of the *Shewanella oneidensis* MR-1 decaheme cytochrome MtrA. J Biol Chem 278:27758–27765
85. Qian Y, Paquete C, Louro R, Ross D, LaBelle E, Bond D, Tien M (2011) Mapping the iron binding site(s) on the small tetraheme cytochrome of *Shewanella oneidensis* MR-1. Biochem 50:6217–6224
86. Reguera G, McCarthy KD, Mehta T, Nicoll JS, Tuominen MT, Lovley DR (2005) Extracellular electron transfer via microbial nanowires. Nature 435:1098–1101
87. Reyes C, Zhang Q, Bondarev S, Welch A, Thelen M, Saltikov C (2012) Characterization of axial and proximal histidine mutations of the decaheme cytochrome MtrA from *Shewanella* sp. strain ANA-3 and implications for the electron transport system. J Bacteriol 194:5840–5847
88. Richardson D, Butt J, Fredrickson J, Zachara J, Shi L, Edwards M, White G, Balden N, Gates A, Marritt S, Clarke T (2012) The ‘porin cytochrome’ model for microbe-to-mineral electron transfer. Mol Microbiol 85:201–212
89. Romine M, Carlson T, Norbeck A, McCue L, Lipton MS (2008) Identification of mobile elements and pseudogenes in the *Shewanella oneidensis* MR-1 genome. Appl Environ Microbiol 74:3257–3265
90. Ross DE, Brantley SL, Tien M (2009) Kinetic characterization of OmcA and MtrC, terminal reductases involved in respiratory electron transfer for dissimilatory iron reduction in *Shewanella oneidensis* MR-1. Appl Environ Microbiol 75:5218–5226
91. Ross DE, Ruebush SS, Brantley SL, Hartshorne RS, Clarke TA, Richardson DJ, Tien M (2007) Characterization of protein-protein interactions involved in iron reduction by *Shewanella oneidensis* MR-1. Appl Environ Microbiol 73:5797–5808
92. Saffarini DA, Schultz R, Beliaev A (2003) Involvement of cyclic AMP (cAMP) and cAMP receptor protein in anaerobic respiration of *Shewanella oneidensis*. J Bacteriol 185:3668–3671
93. Satomi M, Fonnesbech Vogel B, Gram L, Venkateswaran K (2006) *Shewanella hafniensis* sp. nov. and *Shewanella morhuae* sp. nov., isolated from marine fish of the Baltic Sea. Int J Syst Evol Microbiol 56:243–249
94. Schicklberger M, Bücking C, Schuetz B, Heide H, Gescher J (2011) Involvement of the *Shewanella oneidensis* decaheme cytochrome MtrA in the periplasmic stability of the beta-barrel protein MtrB. Appl Environ Microbiol 77:1520–1523
95. Schuetz B, Schicklberger M, Kuermann J, Spormann A, Gescher J (2009) Periplasmic electron transfer via the *c*-type cytochromes MtrA and FccA of *Shewanella oneidensis* MR-1. Appl Environ Microbiol 75:7789–7796
96. Schwalb C, Chapman SK, Reid G (2003) The tetraheme cytochrome CymA is required for anaerobic respiration with dimethyl sulfoxide and nitrite in *Shewanella oneidensis*. Biochemistry 42:9491–9497

97. Schwalb C, Chapman SK, Reid GA (2002) The membrane-bound tetrahaem c-type cytochrome CymA interacts directly with the soluble fumarate reductase in *Shewanella*. Biochem Soc Trans 30:658–662
98. Shi L, Belchik S, Wang Z, Kennedy D, Dohnalkova A, Marshall M, Zachara J, Fredrickson J (2011) Identification and characterization of $UndA_{HRCR\text{-}6}$, an outer memrbane endecaheme *c*-type cytochrome of *Shewanella* sp. Strain HRCR-6. Appl Environ Microbiol 77:5521–5523
99. Shi L, Chen B, Wang Z, Elias DA, Mayer MU, Gorby YA, Ni S, Lower BH, Kennedy DW, Wunschel DS, Mottaz HM, Marshall MJ, Hill EA, Beliaev AS, Zachara JM, Fredrickson JK, Squier TC (2006) Isolation of a high-affinity functional protein complex between OmcA and MtrC: Two outer membrane decaheme *c*-type cytochromes of *Shewanella oneidensis* MR-1. J Bacteriol 188:4705–4714
100. Shi L, Deng S, Marshall MJ, Wang Z, Kennedy DW, Dohnalkova AC, Mottaz HM, Hill EA, Gorby YA, Beliaev AS, Richardson DJ, Zachara JM, Fredrickson JK (2008) Direct involvement of type II secretion system in extracellular translocation of *Shewanella oneidensis* outer membrane cytochromes MtrC and OmcA. J Bacteriol 190:5512–5516
101. Shi L, Rosso K, Clarke T, Richardson D, Zachara J, Fredrickson J (2012) Molecular underpinnings of Fe(III) oxide reduction by *Shewanella oneidensis* MR-1. Front Microbiol 3:1–10
102. Shirodkar S, Reed S, Romine M, Saffarini D (2011) The octaheme SirA catalyses dissimilatory sulfite reduction in *Shewanella oneidensis* MR-1. Environ Microbiol 13:108–115
103. Strzepek R, Maldonado M, Higgins J, Hall J, Safi K, Wilhelm S, Boyd P (2005) Spinning the "ferrous wheel": the importance of the microbial community in an iron budget during the FeCycle experiment. Global Biogeochem Cycles 19:GB4S26
104. Tsapin A, Vandenberghe I, Nealson K, Scott J, Meyer T, Cusanovich M, Harada E, Kaizu T, Akutsu H, Leys D, Van Beeumen J (2001) Identification of a small tetraheme cytochrome *c* and a flavocytochrome *c* as two of the principal soluble cytochromes *c* in *Shewanella oneidensis* strain MR1. Appl Environ Microbiol 67:3236–3244
105. Vargas M, Malvankar N, Tremblay P-L, Leang C, Smith J, Patel P, Snoeyenbos-West O, Nevin K, Lovley D (2013) Aromatic amino acids required for pili conductivity and long-range extracellular electron transport in *Geobacter sulfurreducens*. mBio 4:e00105–e00113
106. Venkateswaran K, Moser DP, Dollhopf ME, Lies DP, Saffarini DA, MacGregor BJ, Ringelberg DB, White DC, Nishijima M, Sano H, Burghardt J, Stackebrandt E, Nealson KH (1999) Polyphasic taxonomy of the genus *Shewanella* and description of *Shewanella oneidensis* sp. nov. Int J Syst Bacteriol 49(2):705–724
107. von Canstein H, Ogawa J, Shimizu S, Lloyd JR (2008) Secretion of flavins by *Shewanella* species and their role in extracellular electron transfer. Appl Environ Microbiol 74:615–623
108. Wee S, Burns J, DiChristina T (2013) Identification of a molecular signature unique to metal-reducing *Gammaproteobacteria*. FEMS Micorbiol Lett 350:90–99
109. White G, Shi Z, Shi L, Wang Z, Dohnalkova A, Marshall M, Fredrickson J, Zachara J, Butt J, Richardson D, Clarke TA (2013) Rapid electron exchange between surface-exposed bacterial cytochromes and Fe(III) minerals. Proc Natl Acad Sci USA 110:6346–6351
110. Wigginton NS, Rosso KM, Hochella MF Jr (2007) Mechanisms of electron transfer in two decaheme cytochromes from a metal-reducing bacterium. J Phys Chem B 111:12857–12864
111. Xiong Y, Chen B, Shi L, Fredrickson J, Bigelow D, Squier TC (2011) Targeted degradation of outer membrane decaheme cytochrome MtrC metal reductase in *Shewanella oneidensis* MR-1 using bioarsenical probe CrAsH-EDT_2. Biochem 50:9738–9751
112. Xiong Y, Shi L, Chen B, Mayer MU, Lower B, Londer Y, Bose S, Hochella MF, Fredrickson J, Squier TC (2006) High-affinity binding and direct electron transfer to solid metals by the *Shewanella oneidensis* MR-1 outer membrane *c*-type cytochrome OmcA. J Am Chem Soc 128:13978–138979
113. Xu M, Guo J, Cen Y, Zhong X, Cao W, Sun G (2005) *Shewanella decolorationis* sp. nov., a dye-decolorizing bacterium isolated from activated sludge of a waste-water treatment plant. Int J Syst Evol Microbiol 55:363–368

114. Yang C, Rodionov DA, Li X, Laikova O, Gelfand M, Zagnitko O, Romine M, Obraztsova A, Nealson K, Osterman A (2006) Comparative genomics and experimental characterization of N-acetylglucosamine utilization pathway of *Shewanella oneidensis*. J Biol Chem 281:29872–79885
115. Yang Y, Chen J, Qiu D, Zhou J (2013) Roles of UndA and MtrC if *Shewanella putrefaciens* W3-18-1 in iron reduction. BMC Microbiol 13:267
116. Zargar K, Saltikov C (2009) Lysine-91 of the tetraheme *c*-type cytochrome CymA is essential for quinone interactions and arsenate respiration in *Shewanella* sp. strain ANA-3. Arch Microbiol 191:797–806
117. Zhang H, Tang X, Munske G, Zakharova N, Yang L, Zheng C, Wolff M, Tolic N, Anderson G, Shi L, Marshall M, Fredrickson J, Bruce J (2008) In vivo identification of the outer membrane protein OmcA-MtrC interaction network in *Shewanella oneidensis* MR-1 cells using novel hydrophobic chemical cross-linkers. J Proteome Res 7:1712–1720

Chapter 3
Collection and Enrichment of Magnetotactic Bacteria from the Environment

Zachery Oestreicher, Steven K. Lower, Dennis A. Bazylinski and Brian H. Lower

Abstract We describe a relatively inexpensive and effective method for collecting magnetotactic bacteria (MTB) from the field. This protocol relies on the use of simple magnets. A clear plastic container can be used to collect sediment and water from a natural source, such as a freshwater pond. In the Northern hemisphere, the south end of a bar magnet is placed against the outside of the container just above the sediment at the sediment-water interface. After some time, the bacteria can be removed from the inside of the container near the magnet with a pipette and then enriched further by using a capillary racetrack and a magnet. In the racetrack, a sterile cotton plug is used to separate magnetic versus non-magnetic cells as the MTB swim through the cotton towards a magnet placed at the opposite end of the racetrack. Once enriched, the presence of MTB can be confirmed by using the hanging drop method and a light microscope to observe MTB swimming in response to the north or south end of a bar magnet. Higher resolution can be obtained by depositing a drop of enriched MTB onto a copper grid and observing the microorganisms with transmission electron microscopy (TEM). Using this method, isolated MTB may be studied microscopically to determine characteristics such as swimming behavior, type and number of flagella, cell morphology, shape of the magnetic crystals, number of magnetosomes, number of magnetosome chains in each cell, composition of the crystals, and presence of intracellular vacuoles.

Keywords Magnetotactic · Magnetite · Magnetosome

Z. Oestreicher · S.K. Lower · B.H. Lower
The Ohio State University, Columbus, OH, USA

D.A. Bazylinski
University of Nevada at Las Vegas, Las Vegas, NV, USA

B.H. Lower (✉)
School of Environment and Natural Resources, 210 Kottman Hall,
2021 Coffey Road, Columbus, OH 43210, USA
e-mail: Lower.30@osu.edu

D. Saffarini (ed.), *Bacteria-Metal Interactions*,
DOI 10.1007/978-3-319-18570-5_3

3.1 Introduction

Magnetotactic bacteria (MTB) are aquatic microorganisms that were first notably described in 1975 from sediment samples collected in salt marshes of Massachusetts (USA) [13]. Since then MTB have been discovered in stratified water and sediment columns from all over the world [14]. One feature common to all MTB is that they contain magnetosomes, which are intracellular, membrane-bound magnetic nanocrystals of magnetite (Fe_3O_4), greigite ($FeSO_4$), or both [7, 36, 42]. In the Northern hemisphere, MTB are typically attracted to the south end of a bar magnet, while in the Southern hemisphere they are usually attracted to the north end of a magnet [7, 58]. This property can be exploited when trying to isolate MTB from environmental samples [48].

The chemical composition, size and morphology of the mineral crystals are under strict control within the magnetosomes [7, 37, 42]. In addition, the arrangement of the magnetosomes within the bacterium is precisely controlled in MTB [7, 37, 42]. These are characteristics of biologically controlled mineralization, as opposed to biologically induced mineralization, where several key steps are under genetic control [5, 9, 12, 35, 42]. Based on the oldest so-called magnetofossils [15, 32, 43, 49, 65], magnetosomes synthesis likely represents the first example of biologically controlled mineralization on Earth [55].

MTB are either obligate microaerophiles or microaerophiles that are facultatively anaerobic or obligate anaerobes. Magnetosomes are most often arranged as a chain within the cell (Fig. 3.1), which cause the cell to passively orient and actively swim along geomagnetic field lines [14, 20, 22, 27]. MTB behave, in effect, as miniature, motile, living compass needles. The original hypothesis for the biologically adaptive value of magnetotaxis was that it enabled magnetotactic bacteria to swim downward along geomagnetic field lines towards less-oxygenated regions of the water column (or sediment) where oxygen and redox conditions, and perhaps nutrient requirements, are most favorable [14, 18, 20, 22, 27, 42]. Recent discoveries that local molecular oxygen and/or hydrogen sulfide concentrations control swimming direction of at least one MTB species [20] and regulate the biomineralization process [7], suggest the possibility that magnetite-producing MTB use aerotaxis (chemotaxis in response to oxygen) in conjunction with magnetotaxis to locate and maintain their optimal position within a water column or sediment [7, 20, 22, 25, 26, 62].

The mineral synthesized in the bacterial magnetosome appears to be species-specific as bacteria usually biomineralize either iron oxide crystals of Fe_3O_4 or iron sulfide crystals of Fe_3S_4 [3, 4, 10–12, 45, 60, 61]. Rarely does a bacterium synthesize both minerals, although there are some exceptions (e.g., see [4, 10, 38]).

Iron oxide magnetosomes consist of stoichiometric magnetite and recent evidence suggests that magnetite forms through phase transformation from a highly disordered phosphate-rich ferric hydroxide phase, consistent with prokaryotic ferritins, through transient nanometric ferric (oxyhydr)oxide intermediates within the magnetosome organelle [2]. In a magnetic and structural study of

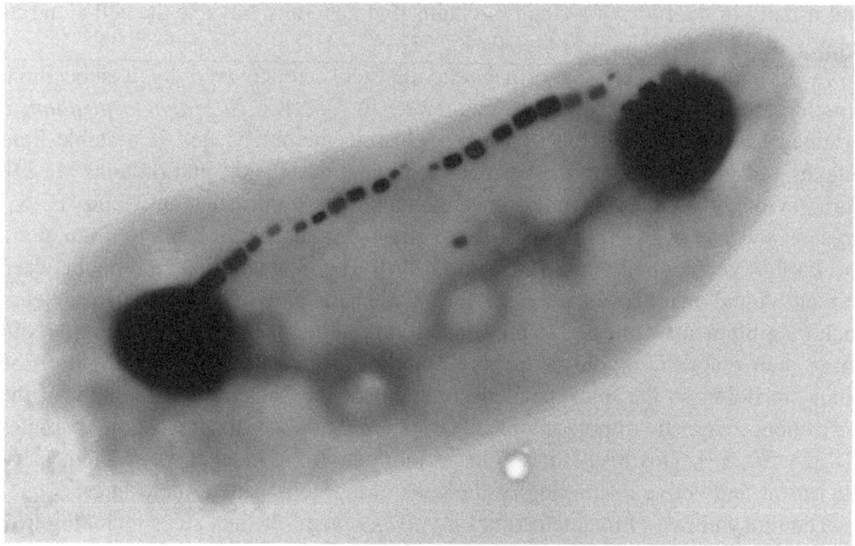

Fig. 3.1 Transmission electron microscope image of a magnetotactic bacterium isolated from the wetland near the Olentangy River in Columbus, Ohio

magnetosomes of *M. gryphiswaldense* using a combination of iron K-edge X-ray absorption near edge structure (XANES) and high-resolution TEM, two phases of iron, ferrihydrite and magnetite, were identified and quantified [19] suggesting that ferrihydrite is the source of iron ions for magnetite biomineralization in *Magnetospirillum gryphiswaldense*. The magnetite biomineralization process is thought to proceed in two steps: first, iron appears to accumulate in the form of ferrihydrite; and second, magnetite rapidly biomineralizes from ferrihydrite. XANES analysis suggests the origin of the ferrihydrite is bacterial ferritin cores characterized by high phosphorus content and a poorly crystalline structure [19].

Magnetite magnetosome crystals are of high chemical purity with few impurities [3, 11, 16, 64]. Iron sulfide magnetosomes contain greigite (Fe_3S_4) or a mixture of greigite and non-magnetite iron sulfide mineral phases including mackinawite (tetragonal FeS) or sphalerite-like, cubic FeS, which appear to be precursor phases for greigite [51–53]. The morphology of Fe_3O_4 and Fe_3S_4 crystals varies from species to species, but is highly conserved within the same bacterial species and in one case, perhaps, within a genus (i.e., *Magnetospirillum*) [3, 10, 17]. Three common crystal morphologies have been described in all MTB based on electron microscopy of the crystal structures: (i) cuboctahedral [1, 31, 44], (ii) elongated prismatic [8, 31, 47], and (iii) bullet-shaped [6, 10, 17, 36, 37, 63]. The size of magnetosome crystals range from about 35 to 120 nm in diameter, and appears to be under strict biological control as all magnetosome crystals, regardless of whether they consist of magnetite or greigite, are single-domain magnets [10, 21, 30, 56]. Each species or strain exhibits a particular arrangement of magnetosomes within the cell usually in the form of a chain or chains that generally traverse the cell along its long axis (Figs. 3.1 and 3.5) [4, 10, 45, 61].

In all cultured MTB, nano-magnetic particles are covered by a subcellular structure called the magnetosome membrane [6, 35, 42]. In *Magnetospirillum*, it originates as an invagination of the cytoplasmic membrane and is a stable lipid bilayer 3–4 nm thick comprised of phospholipids, fatty acids, and proteins [1, 29]. As previously stated, magnetosomes contain single-domain Fe_3O_4 (or Fe_3S_4) crystals and are arranged in one or more chains that cause the cells to align along the Earth's geomagnetic field [21, 30, 45, 56]. The magnetic interactions between the individual magnetosome crystals in the chain and a complex cytoskeleton including filaments dedicated to the construction of the magnetosome chain [34, 59] cause their magnetic dipole moments to orient head-to-tail along the length of the chain. In doing so the total magnetic moment of the bacterium is the sum of the permanent magnetic dipole moments of the individual magnetosome particles [18, 28, 50, 54]. This results in the bacterium exhibiting magnetotaxis, the passive alignment and active swimming of the cells along geomagnetic field lines.

The biosynthesis of magnetosomes involves several distinct steps including iron uptake by the cell, magnetosome vesicle formation, iron transport into the magnetosome vesicle, and protein-mediated Fe_3O_4 or Fe_3S_4 biomineralization within the magnetosome vesicle [6, 21, 23, 24]. A number of proteins located on or in the magnetosome membrane have been isolated from MTB (Table 3.1). All proteins in Table 3.1 are from magnetite-producing MTB because these strains, as opposed to greigite-producers, are "relatively" easy to grow in pure culture. The so-called Mms proteins (or their homologues; e.g., MamC, D, G) are particularly interesting because they seem to be critical for the nucleation, growth and maturity of magnetite crystals (Table 3.1). These Mms proteins share similar features in their primary amino acid sequences. Each of the Mms proteins contains hydrophobic N-terminus, which may serve as a transmembrane domain that anchors the proteins into the lipid bilayer membrane of the magnetosome vesicle. Each Mms protein also

Table 3.1 Magnetosome proteins from *M. magneticum* AMB-1

Magnetosome protein	Putative function
MamA, Mam22, Mam24	Scaffold proteins, coordinate assembly of protein complexes
MamB, MamM, MamN, MamV	Transport Fe into magnetosome, form protein-protein complexes
MamC/Mms13, MamD/Mms7, MamG/Mms5, Mms6	Mineral nucleation, crystal-lization, mineral size and shape
MamE, MamO, MamP	Control arrangement of proteins within magnetosome membrane
MamI, MamL, MamQ	Magnetosome membrane invagination, bilayer formation
MamJ, MamK	Assembly of magnetosome chain

Homologous proteins from different MTB are separated by a slash (e.g., MamC is from *M. gryphiswaldense* and Mms13 is from *M. magneticum* AMB-1)

contains a hydrophilic C-terminal region consisting of several highly conserved amino acids that have carboxyl and hydroxyl side groups. Regardless of whether ferrihydrite is an intermediate in magnetite synthesis [19], it is believed that these amino acids function as a template that controls the morphology of the nascent Fe_3O_4 crystals by inhibiting growth in one direction and/or promoting growth in another.

3.2 Magnetotactic Bacteria Collection

When deciding on a freshwater site to collect magnetotactic bacteria (MTB), it is often best to start with a pond or slow-moving stream that has a soft muddy sediment layer. For this isolation protocol, we collected water-sediment samples from a wetland near the campus of The Ohio State University (OSU) in Columbus, Ohio (USA). The protocol described herein can be applied to virtually any aquatic location. The materials used in this protocol can be found in Table 3.2.

Find a location where the depth of the water is between 10 and 100 cm. At such a location, you should collect the upper-most layer of sediment using a clear, screw-top container. Scoop the sediment and water into the container until it is filled with one-third to one-half sediment and the remaining volume with water (Fig. 3.2a). Keep the container submerged until it is filled with water and then tightly cap the container with its screw-top lid. It's not necessary to mix the sediment. Wipe the outside of the container dry with a towel and then take the sample to your laboratory. It is not necessary to rush the sample back to your laboratory. The MTB should be viable for several weeks to months as long as you store the samples in a cool, dark location with the cap loosened.

Once the sample is in your laboratory, loosen the cap but leave it covering the container to reduce the amount of evaporation. Store the container at room temperature in a dark room, drawer, or completely cover the container with aluminum foil. Allow the sediment and fine particles to completely settle to the bottom of the container by leaving the sample undisturbed for several hours to several days. It is not necessary to mix the sediment, MTB prefer an undisturbed environment. The clear walls of the plastic container will allow you to confirm that the particles have settled to the bottom. Depending on your sample, MTB can remain alive in the sample for many months.

3.3 Magnetotactic Bacteria Isolation

When you are ready to isolate the MTB, place magnets on the sides of the plastic container approximately 1 cm above the sediment-water interface (Fig. 3.2). Be careful not to disturb the sediment in the bottom of the container. Place the south pole of a bar magnet on one side of the container and the north side of another bar

Table 3.2 List of specific reagents and equipment used to isolate MTB and study the microorganisms

Item name	Company	Catalogue number	Comments
0.22 μm filter	Fisher Scientific	09-719C	
1 mL syringe	Fisher Scientific	NC9788564	
Bar magnet	Fisher Scientific	S95957	
Container	Any		Any plastic or glass container that can hold at least 0.5 L and can be sealed
Cotton	Any		
Cover slips	Fisher Scientific	12-542B	
Diamond pen	Fisher Scientific	08-675	
Formvar/Carbon 200 mesh, copper grids	Ted Pella Inc.	01800	
Glass Pasteur pipets	Fisher Scientific	13-678-6A	
Glass slides	Fisher Scientific	S95933	
Microcentrifuge tubes	Fisher Scientific	02-681-320	
Microscope with 60× dry lens	Zeiss		A 60× dry lens is not absolutely necessary, but this gives a high NA without using oil
O-ring	Hardware store		
Tecnai F20 S/TEM	FEI		
Tecnai Spirit TEM	FEI		
Uranyl acetate	Ted Pella Inc.	19481	

magnet on the opposite side (Fig. 3.2). Almost any magnet can be used, such as a magnetic stir bar or large refrigerator magnet. Anything can be used to support the magnets at the correct height above the sediment-water interface. We have found that resting the magnets on the top of a cardboard or plastic box is best, however, the magnets can also be taped to the outside of the plastic container. Wait 30 min to several hours for the bacteria to swim to the magnet.

Use a sterile pipette to carefully collect the water from inside the container (Fig. 3.2) near the position of the south pole of the bar magnet (for samples collected in the Northern hemisphere). This water should contain the MTB that have been attracted to the south pole of the bar magnet. Next, a capillary racetrack should be used to further enrich the MTB.

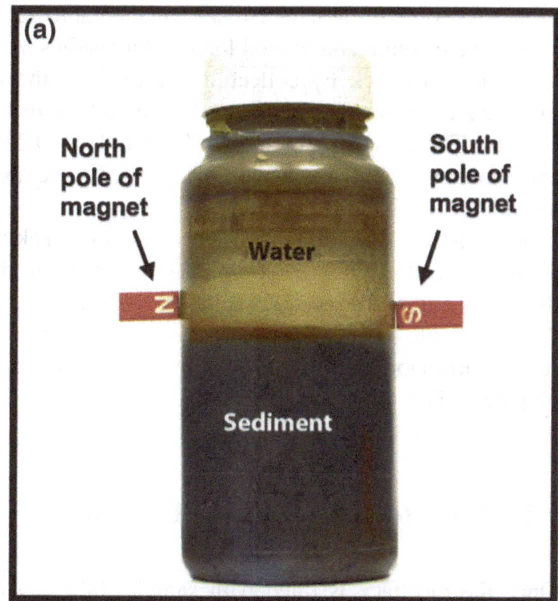

Fig. 3.2 A clear plastic bottle (1-L) containing a sediment and water sample collected from the Olentangy River in Columbus, Ohio (USA). The bottle contains approximately one-half sediment and one-half water. The south end of a magnet is placed approximately 1 cm above the sediment for up to several hours (**a**). After removing some of the fluid from near the magnet on the inside the container, it is placed inside of a capillary racetrack where the MTB swim through a cotton plug (*arrow*) towards the south end of a bar magnet (**b**). A close up view of the capillary racetrack showing the sample, cotton, filtered fluid, sealed end of the capillary tube and south end of a bar magnet (**c**)

3.4 Magnetotactic Bacteria Racetrack

In order to enrich your sample with MTB, a capillary racetrack is necessary (Fig. 3.1b). These need to be made prior to isolating the cells from the clear-plastic container. Use a 5.75 in. (146 mm) glass Pasteur pipette to make a racetrack. Use a diamond pen or file to cut off the top of the pipette, the length of the pipette is not crucial, but it should be able to contain approximately 1–2 mL of water. Next, use a Bunsen burner to melt the tip so that it becomes sealed (Fig. 3.2b). The resulting pipette should have an open end and a sealed end (Fig. 3.2b).

Make several of these racetracks and then autoclave. Additionally, you will need to autoclave cotton and several long metal needles. Add filtered sample water to the tip of the racetrack by collecting liquid from the top of the sample shown in Fig. 3.2a, to an autoclaved racetrack using a long metal needle attached to a filtered syringe. The pore size of the filter should be 0.22 μm to eliminate debris and contaminants from the water. It is important to be absolutely sure that there are no air bubbles in the glass capillary.

Plug the neck of the sealed half of the racetrack with sterile cotton (Fig. 3.2b) about 0.5 cm from the sealed tip. Use the metal needle to push the cotton towards the sealed end of the racetrack so it is 0.5–1 cm away from the sealed tip (Fig. 3.2c). Using a sterile pipette, remove MTB-containing fluid (as described in the previous section) from the sample container and add it to the sample reservoir (open end) of a prepared MTB racetrack (Fig. 3.2b).

3.5 Magnetotactic Bacteria Enrichment

Once the racetrack is filled with sample fluid, lay it on its side on a horizontal surface (e.g., a benchtop) and place the south pole of a bar magnet (in the Northern hemisphere) next to the sealed tip of the racetrack (Fig. 3.2). Wait 5–30 min for the MTB to migrate through the cotton. Then you should collect the fluid near the tip of the racetrack. Waiting too long can introduce contaminants, such as other motile bacteria, to the tip of the capillary. Optionally, you could use a light microscope to view the tip of the racetrack and watch the MTB collect at the racetrack's tip. This will allow you to determine how long it takes the MTB to migrate through the cotton plug.

Then to remove the enriched MTB, gently use the diamond knife to make a little scratch near the cotton plug and snap off the end of the racetrack. Use a 1 mL syringe with a narrow needle (25 or 27 gauge) to remove the fluid from the tip of the racetrack. This liquid sample should now contain the enriched MTB.

3.6 Magnetotactic Bacteria Observation by Light Microscopy

Place a drop (10–20 μL) of the enriched MTB sample onto a coverslip. Quickly flip the coverslip over so the drop is now facing down and hanging from the coverslip (Fig. 3.3). Place the coverslip onto an O-ring that is resting on a glass slide (Fig. 3.3). The O-ring should have a slightly smaller diameter then the coverslip (about 1 cm; Fig. 3.3). Place this hanging drop onto a light microscope stage and focus on one edge of the drop. A 60× dry objective works very well because most

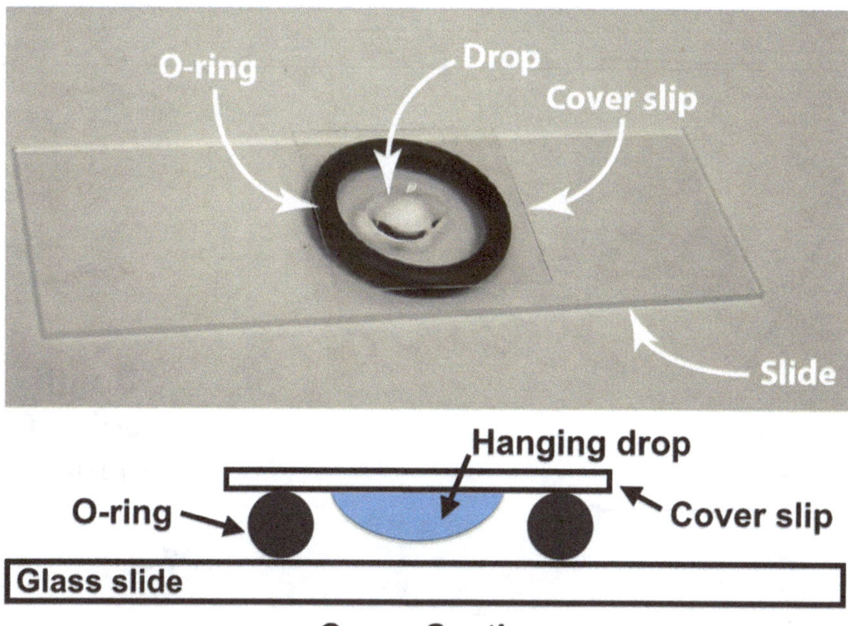

Fig. 3.3 Once the MTB have been enriched from the racetrack, a small drop can be placed on a coverslip, which is then turned upside down and placed on an O-ring that is sitting on a slide. This slide-O-ring-coverslip sandwich can be placed on a light microscope stage and viewed using a 60× dry objective (oil lenses are inconvenient to use with this method)

have a high numerical aperture, but do not require oil, which is difficult to use when using the hanging drop method (Fig. 3.3). Place the south end of a bar magnet close to the hanging drop and MTB will begin to migrate towards the edge of the drop closest to the magnet (Fig. 3.4). Within a few minutes many MTB should be at the edge of the drop (Fig. 3.4). You are able to prove that the bacteria are magnetic by reversing the pole of the magnet and then observe the bacteria swim in the opposite direction.

3.7 Magnetotactic bacteria Observation by Transmission Electron Microscopy (TEM)

Place a drop (~20 μL) of the enriched MTB onto a flat surface, such as parafilm, and place a copper grid on the drop, and allow the bacteria to adsorb onto the grid for about 10–20 min. Wick off excess water with a piece of clean filter paper.

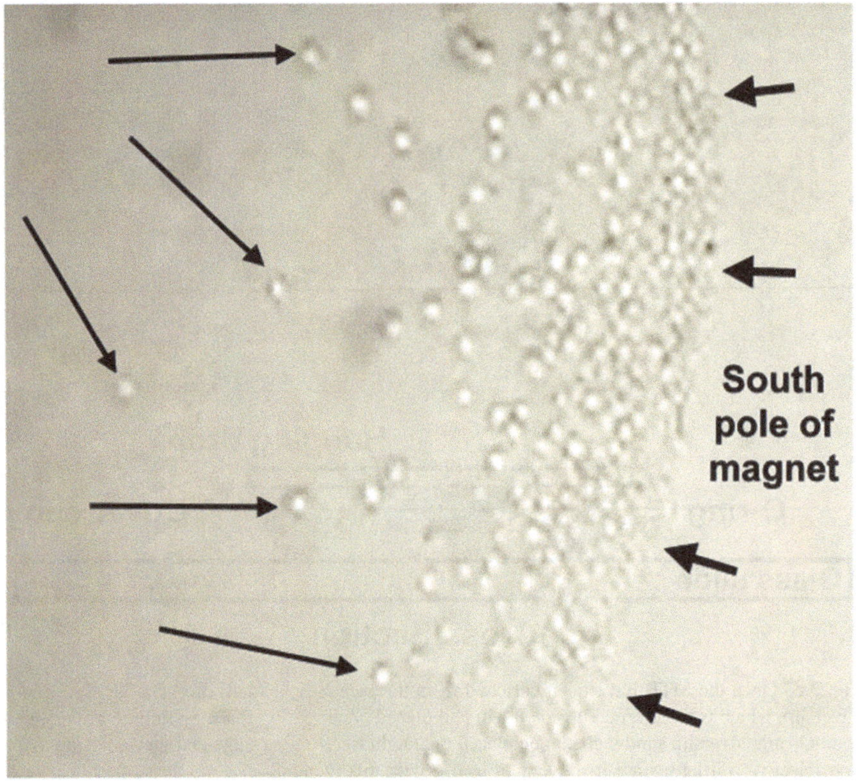

Fig. 3.4 Bright field microscope image of MTB swimming (*thin long arrows*) and gathering at the edge of the hanging drop (*short arrows*) which is next to the south pole of a bar magnet

Optionally, the grid can be negatively stained with 2 % uranyl acetate, 2 % phosphotungstic acid pH 7.2, or 2.5 % sodium molybdate [1, 46, 57]. This is done by placing the copper grid onto a drop of stain immediately after incubating the grid with the enriched MTB. Incubate the grid with the negative stain; the times will vary depending on the stain used, and then wick off the fluid with a piece of clean filter paper. Observe the MTB using transmission electron microscopy (TEM; Fig. 3.5). For the work described here MTB were adsorbed to Formvar stabilized and carbon coated 200 mesh copper grids (Ted Pella #01800). The grids were placed with the carbon side down on a drop of cell suspension for up to 10 min, then immediately washed one time by placing the grid on a drop of water for 30 s. For staining, the grids were placed on a drop of 2 % uranyl acetate (Ted Pella #19481) for 30 s to 5 min and then dried completely using a piece of filter paper. The grids were analyzed by TEM using either an FEI Tecnai Spirit at 80 kV.

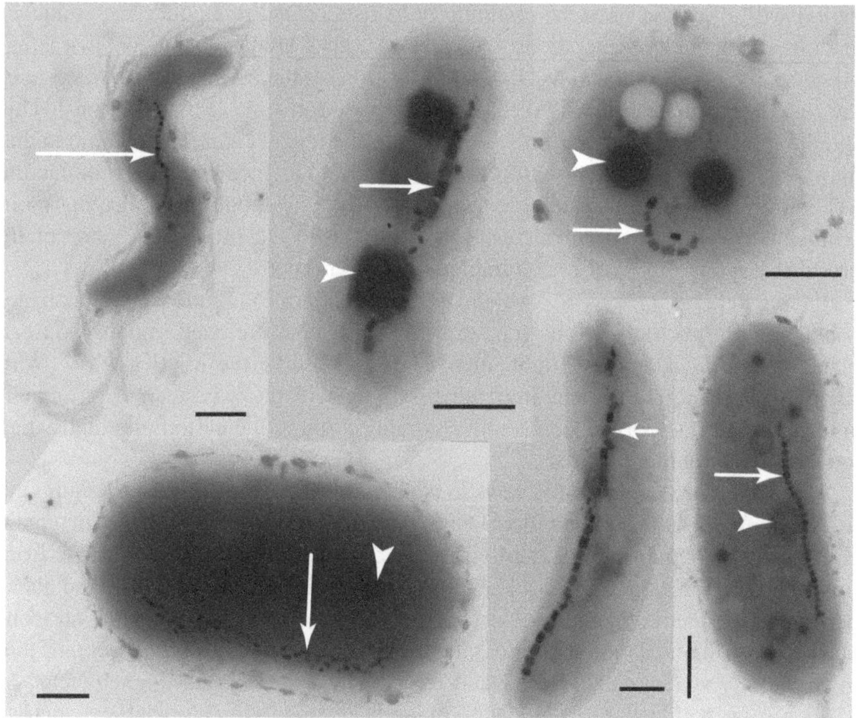

Fig. 3.5 Transmission electron microscope images of various MTB isolated from the wetland near the Olentangy River in Columbus, Ohio. There were several different morphotypes, all containing crystals (*white arrows*). Some of the microorganisms contained inclusions (*white arrowheads*). Scale bar for each image is 500 nm

3.8 Discussion

Magnetotactic bacteria are not necessarily found in every aquatic environment but when they do occur, there can be as many as 100–1000 cells/mL [14, 46]. In order to observe the MTB using optical microscopy, you will need approximately 50 bacteria/mL in your sample [46]. If there are no or few MTB in your sample, then you will need to either select a new environmental site to collect your samples or try an enrichment technique like the one described here. First, a magnet is used to isolate or concentrate magnetotactic bacteria (MTB) contained in the environmental sample (Fig. 3.2a). Then a capillary racetrack (Fig. 3.2b) can be used to attract MTB through a cotton plug where they can be separated from non-magnetotactic microorganisms also contained within the environmental sample.

When you bring your environmental sample back to the laboratory, it may be beneficial to wait several days or weeks for the sample to acclimate to laboratory conditions before trying to isolate the MTB using a bar magnet. This acclimation period will allow the bacterial community to mature and repopulate the container leading to higher concentrations of MTB. Another simple technique that often produces more concentrated MTB samples is to leave the bar magnet on the side of the sample container (Fig. 3.2a) for a longer period of time (e.g., overnight). This should allow the MTB more time to migrate to the magnet. Another technique that may be useful, is to use several racetracks (Fig. 3.2b) at once and then combine the MTB from each racetrack into one sample. Lastly, you should try collecting more sediment from the environment using a large plastic tub [46]. This is especially useful if large numbers of unculturable MTB are needed.

If you believe there is a problem with a racetrack or if there are too many contaminating microorganisms (i.e., non-MTB) in your enriched sample, you can place the racetrack under a light microscope to observe the MTB as they swim through the cotton plug and into the tip. This will allow you to determine if contaminating microorganisms are also coming through the cotton plug and when to stop the enrichment process.

There are more sophisticated ways to isolate MTB, but these methods require the use of more specialized equipment. One example involves the use of a magnetic coil, instead of a bar magnet, and customized glass vessels to isolate MTB from freshwater sediments [33, 41]. The protocol described here represents an inexpensive and effective method for determining whether an environmental site contains MTB. This isolation and enrichment protocol is straightforward enough that microbiology students can master and easily "fine-tune" so that higher yields of MTB can be achieved. Once the MTB have been isolated, other analyses such as fluorescence in situ hybridization, 16S rRNA sequencing for community analysis, energy dispersive spectroscopy (EDS), TEM, optical microscopy, and magnetic measurements can be conducted on the MTB [39, 40].

References

1. Balkwill DL, Maratea D, Blakemore RP (1980) Ultrastructure of a magnetotactic spirillum. J Bacteriol 141:1399–1408
2. Baumgartner J, Morin G, Menguy N, Perez Gonzalez T, Widdrat M, Cosmidis J, Faivre D (2013) Magnetotactic bacteria form magnetite from a phosphate-rich ferric hydroxide via nanometric ferric (oxyhydr)oxide intermediates. Proc Natl Acad Sci USA 110:14883–14888
3. Bazylinski DA (1996) Controlled biomineralization of magnetic minerals by magnetotactic bacteria. Chem Geol 132:191–198
4. Bazylinski DA (1995) Structure and function of the bacterial magnetosome. ASM News 61:337–343
5. Bazylinski DA, Frankel RB (2003) Biologically controlled mineralization in prokaryotes. In: Dove PM, DeYoreo JJ, Weiner S (eds) Reviews in mineralogy and geochemistry—biomineralization, vol 54. Mineralogical Society of America, Geochemical Society, pp 217–247
6. Bazylinski DA, Frankel RB (2004) Magnetosome formation in prokaryotes. Nat Rev Microbiol 2:217–230

7. Bazylinski DA, Frankel RB, Heywood BR, Mann S, King JW, Donaghay PL, Hanson AK (1995) Controlled biomineralization of magnetite (Fe_3O_4) and greigite (Fe_3S_4) in magnetotactic bacterium. Appl Environ Microbiol 61:3232–3239
8. Bazylinski DA, Frankel RB, Jannasch HW (1988) Anaerobic magnetite production by a marine, magnetotactic bacterium. Nature 334:518–519
9. Bazylinski DA, Frankel RB, Konhauser KO (2007) Modes of biomineralization of magnetite by microbes. Geomicrobiol J 24:465–475
10. Bazylinski DA, Garratt-Reed AJ, Frankel RB (1994) Electron microscopic studies of magnetosomes in magnetotactic bacteria. Microsc Res Tech 27:389–401
11. Bazylinski DA, Garratt-Reed DA, Abedi AJ, Frankel RB (1993) Copper association with iron sulfide magnetosomes in a magnetotactic bacterium. Arch Microbiol 160:35–42
12. Bazylinski DA, Schubbe S (2007) Controlled biomineralization by and applications of magnetotactic bacteria. Adv Appl Microbiol 62:21–62
13. Blakemore R (1975) Magnetotactic bacteria. Science 190(4212):377–379
14. Blakemore RP (1982) Magnetotactic bacteria. Annu Rev Microbiol 36:217–238
15. Chang S-BR, Kirschvink JL (1989) Magnetofossils, the magnetization of sediments, and the evolution of magnetite biomineralization. Annu Rev Earth Planet Sci 17:169–195
16. Clemett SJ, Thomas-Keprta KL, Shimmin J, Morphew M, McIntosh JR, Bazylinski DA, Kirschvink JL, Wentworth SJ, McKay DS, Vali H, Gibson EK, Romanek CS (2002) Crystal morphology of MV-1 magnetite. Am Mineral 87:1727–1730
17. Devouard B, Posfai M, Hua X, Bazylinski DA, Frankel RB, Buseck PR (1998) Magnetite from magnetotactic bacteria: size distributions and twinning. Am Mineral 83:1387–1398
18. Dunin-Borkowski RE, McCartney MR, Frankel RB, Bazylinski DA, Posfai M, Buseck PR (1998) Magnetic microstructure of magnetotactic bacteria by electron holography. Science 282:1868–1870
19. Fdez-Gubieda ML, Muela A, Alonso J, Garcia-Prieto A, Olivi L, Fernandez-Pacheco R, Barandiaran JM (2013) Magnetite biomineralization in *Magnetospirillum gryphiswaldense*: time-resolved magnetic and structural studies. ACS Nano 7:3297–3305
20. Frankel R, Bazylinski D, Johnson M, Taylor B (1997) Magneto-aerotaxis in marine coccoid bacteria. Biophys J 73:994–1000
21. Frankel R, Bazylinski D, Schuler D (1998) Biomineralization of magnetic iron minerals in bacteria. Supramol Sci 5:383–390
22. Frankel RB (1984) Magnetic guidance of organisms. Annu Rev Biophys Bioeng 13:85–103
23. Frankel RB, Bazylinski DA (2003) Magnetosome mysteries. ASM News 70:176–183
24. Frankel RB, Bazylinski DA (2003) Magnetosomes: nanoscale magnetic iron minerals in bacteria. In: Niemeyer CM, Mirkin C (eds) Nanobiotechnology: concepts, applications and perspectives. Wiley-VCH, Weinheim, pp 136–145
25. Frankel RB, Bazylinski DA (2009) Magnetosomes and magneto-aerotaxis. In Collin M, Schuch R (eds) Bacterial sensing and signaling, vol 16, pp 182–193
26. Frankel RB, Bazylinski DA, Johnson MS, Taylor BL (1997) Magneto-aerotaxis in marine coccoid bacteria. Biophys J 73:994–1000
27. Frankel RB, Williams TJ, Bazylinski DA (2006) Magneto-aerotaxis. In: Schüler D (ed) Magnetoreception and magnetosomes in bacteria. Springer, Berlin, pp 1–24
28. Frankel RB, Zhang JP, Bazylinski DA (1998) Single magnetic domains in magnetotactic bacteria. J Geophys Res Solid Earth 103:30601–30604
29. Gorby YA, Beveridge TJ, Blakemore RP (1988) Characterization of the bacterial magnetosome membrane. J Bacteriol 170:834–841
30. Hanzlik M, Winklhofer M, Petersen N (1996) Spatial arrangement of chains of magnetosomes in magnetotactic bacteria. Earth Planet Sci Lett 145:125–134
31. Heywood BR, Bazylinski DA, Garrattreed A, Mann S, Frankel RB (1990) Controlled biosyntheisis of greigite (Fe3S4) in magnetotactic bacteria. Naturwissenschaften 77:536–538
32. Jimenez-Lopez C, Romanek CS, Bazylinski DA (2010) Magnetite as a prokaryotic biomarker: a review. J Geophys Res-Biogeosci 115

33. Jogler C, Lin W, Meyerdierks A, Kube M, Katzmann E, Flies C, Pan Y, Amann R, Reinhardt R, Schüler D (2009) Toward cloning of the magnetotactic metagenome: identification of magnetosome island gene clusters in uncultivated magnetotactic bacteria from different aquatic sediments. Appl Environ Microbiol 75(12):3972–3979
34. Komeili A, Li Z, Newman DK, Jensen GJ (2006) Magnetosomes are cell membrane invaginations organized by the actin-like protein MamK. Science 311:242–245
35. Lefèvre CT, Bazylinski DA (2013) Ecology, diversity, and evolution of magnetotactic bacteria. Microbiol Mol Biol Rev 77:497–526
36. Lefèvre CT, Frankel RB, Abreu F, Lins U, Bazylinski DA (2011) Culture-independent characterization of a novel, uncultivated magnetotactic member of the Nitrospirae phylum. Environ Microbiol 13:538–549
37. Lefèvre CT, Menguy N, Abreu F, Lins U, Posfai M, Prozorov T, Pignol D, Frankel RB, Bazylinski DA (2011) A cultured greigite-producing magnetotactic bacterium in a novel group of sulfate-reducing bacteria. Science 334:1720–1723
38. Lefèvre CT, Posfai M, Abreu F, Lins U, Frankel RB, Bazylinski DA (2011) Morphological features of elongated-anisotropic magnetosome crystals in magnetotactic bacteria of the Nitrospirae phylum and the Deltaproteobacteria class. Earth Planet Sci Lett 312:194–200
39. Li J, Pan Y, Liu Q, Yu-Zhang K, Menguy N, Che R, Qin H, Lin W, Wu W, Petersen N, Yang X (2010) Biomineralization, crystallography and magnetic properties of bullet-shaped magnetite magnetosomes in giant rod magnetotactic bacteria. Earth Planet Sci Lett 293(3–4):368–376
40. Lin W, Li J, Pan Y (2012) Newly isolated but uncultivated magnetotactic bacterium of the phylum Nitrospirae from Beijing, China. Appl Environ Microbiol 78(3):668–675
41. Lins U, Freitas F, Keim CN, Barros HL, Esquivel DMS, Farina M (2003) Simple homemade apparatus for harvesting uncultured magnetotactic microorganisms. Braz J Microbiol 34 (2):111–116
42. Lower BH, Bazylinski DA (2013) The bacterial magnetosome: a unique prokaryotic organelle. J Mol Microbiol Biotech 23:63–80
43. Mandernack KW, Bazylinski DA, Shanks WC, Bullen TD (1999) Oxygen and iron isotope studies of magnetite produced by magnetotactic bacteria. Science 285:1892–1896
44. Mann S, Frankel RB, Blakemore RP (1984) Structure, morphology and crystal-growth of bacterial magnetite. Nature 310:405–407
45. Matsunaga T, Sakaguchi T (2000) Molecular mechanism of magnet formation in bacteria. J Biosci Bioeng 90:1–13
46. Moench TT, Konetzka W (1978) A novel method for the isolation and study of a magnetotactic bacterium. Arch Microbiol 119(2):203–212
47. Moench TT (1988) *Bilophococcus-magnetotacticus* Gen-Nov-Sp-No, a motile, magnetic coccus. Antonie Van Leeuwenhoek J Microbiol 54:483–496
48. Oestreicher Z, Lower SK, Lin W, Lower BH (2012) Collection, isolation and enrichment of naturally occurring magnetotactic bacteria from the environment. J Visualized Exp (69): e50123
49. Peck JA, King JW (1996) Magnetofossils in the sediment of Lake Baikal, Siberia. Earth Planet Sci Lett 140:159–172
50. Penninga I, Dewaard H, Moskowitz BM, Bazylinski DA, Frankel RB (1995) Remanence measurements on individual magnetotactic bacteria using a pulsed magnetic field. J Magn Magn Mater 149:279–286
51. Posfai M, Buseck PR, Bazylinski DA, Frankel RB (1998) Iron sulfides from magnetotactic bacteria: structure, composition, and phase transitions. Am Mineral 83:1469–1481
52. Posfai M, Buseck PR, Bazylinski DA, Frankel RB (1998) Reaction sequence of iron sulfide minerals in bacteria and their use as biomarkers. Science 280:880–883
53. Posfai M, Lefevre CT, Trubitsyn D, Bazylinski DA, Frankel RB (2013) Phylogenetic significance of composition and crystal morphology of magnetosome minerals. Front Microbiol 4:344

54. Proksch RB, Schaffer TE, Moskowitz BM, Dahlberg ED, Bazylinski DA, Frankel RB (1995) Magnetic force microscopy of the submicron magnetic assembly in a magnetotactic bacterium. Appl Phys Lett 66:2582–2584
55. Prozorov T, Bazylinski DA, Mallapragada SK, Prozorov R (2013) Novel magnetic nanomaterials inspired by magnetotactic bacteria: topical review. Mat Sci Eng R 74:133–172
56. Ricci JCD, Kirschvink JL (1992) Magnetic domain state and coercivity predictions for biogenic greigite (Fe_3S_4)—a comparison of theory with magnetosome observations. J Geophys Res Solid Earth 97:17309–17315
57. Rodgers FG, Blakemore RP, Blakemore NA, Frankel RB, Bazylinski DA, Maratea D, Rodgers C (1990) Intercellular structure in a many-celled magnetotactic prokaryote. Arch Microbiol 154(1):18–22
58. Simmons SL, Bazylinski DA, Edwards KJ (2006) South-seeking magnetotactic bacteria in the Northern Hemisphere. Science 311(5759):371–374
59. Scheffel A, Gruska M, Faivre D, Linaroudis A, Plitzko JM, Schuler D (2006) An acidic protein aligns magnetosomes along a filamentous structure in magnetotactic bacteria. Nature 440:110–114
60. Spring S, Amann R, Ludwig W, Schleifer KH, Vangemerden H, Petersen N (1993) Dominating role of an unusual magnetotactic bacterium in the microaerobic zone of a freshwater sediment. Appl Environ Microbiol 59:2397–2403
61. Spring S, Schleifer KH (1995) Diversity of magnetotactic bacteria. Syst Appl Microbiol 18:147–153
62. Taylor BL, Zhulin IB, Johnson MS (1999) Aerotaxis and other energy-sensing behavior in bacteria. Annu Rev Microbiol 53:103–128
63. Thornhill RH, Burges JG, Sakaguchi T, Matsunaga T (1994) A morphological classification of bacteria containing bullet-shaped magnetic particles. FEMS Microbiol Lett 115:169–176
64. Towe KM, Moench TT (1981) Electron-optical characterization of bacterial magnetite. Earth Planet Sci Lett 52:213–220
65. Weiss BP, Kim SS, Kirschvink JL, Kopp RE, Sankaran M, Kobayashi A, Komeili A (2004) Ferromagnetic resonance and low-temperature magnetic tests for biogenic magnetite. Earth Planet Sci Lett 224:73–89

Chapter 4
Metabolism of Metals and Metalloids by the Sulfate-Reducing Bacteria

Larry L. Barton, Francisco A. Tomei-Torres, Hufang Xu and Thomas Zocco

Abstract The bacteria and archaea that reduce sulfate to sulfide can transform a variety of metal(loids). The latter include metalloids (As, Se and Te), transition metals (Au, Co, Cr, Fe, Hg, Mo, Mn, Ni, Pb, Pd, Pt, Re, Rh, Tc, V, and Zn), and actinides (Pu and U). The conversions are achieved via (1) use of metal-specific enzymes, (2) cometabolism, i.e., use of non-substrate-specific enzymes, (3) biomethylation, (4) inorganic precipitation, (5) oxidation-reduction reactions in the growth medium; or (6) oxidation/cathodic depolarization of the elemental form. Respective examples are (1) the respiration of arsenate by *Desulfosporosinus auripigmenti*; (2) reduction of selenate and selenite to elemental selenium by enzymes involved in sulfate respiration or assimilation; (3) methylation of mercury; (4) precipitation of zinc sulfide in the supernatant; (5) reaction of sulfide and selenite forming selenium sulfide (SeS_2) in the supernatant; and (6) the anaerobic corrosion of iron. Some of these processes yield valuable commodities, e.g., the precipitation of gold by *Desulfovibrio desulfuricans*. The understanding of anaerobic corrosion can lead to the prevention of corrosion of pipelines. The formation of selenium nanoparticles has potential applications in the design of drug-delivery

L.L. Barton (✉)
Department of Biology, University of New Mexico, MSCO3 2020, Albuquerque, NM 87131-0001, USA
e-mail: lbarton@unm.edu

F.A. Tomei-Torres
Division of Toxicology and Human Health Sciences, Agency for Toxic Substances and Disease Registry, 4770 Buford Highway, NE (Mail Stop F-62), Atlanta, GA 30341-3717, USA
e-mail: fbt3@cdc.gov

H. Xu
Department of Geoscience, University of Wisconsin, Madison, WI 53700, USA
e-mail: hfxu@geology.wisc.edu

T. Zocco
Materials Science Division, Los Alamos National Laboratory, Los Alamos, NM 87545, USA
e-mail: zocco@lanl.gov

D. Saffarini (ed.), *Bacteria-Metal Interactions*,
DOI 10.1007/978-3-319-18570-5_4

systems. The formation of insoluble precipitates facilitates the design of bioremediation technologies. While some metals, e.g., Fe, Co, Mo, Mn, Ni, Se, V and Zn, are essential nutrients for bacterial growth, this review focuses on detoxification processes and not on trace metal assimilation into cellular materials.

Keywords Metal reduction · Anaerobic sulfate respiration · *Desulfovibrio* · Detoxification of metals · Radionuclides

4.1 Introduction

The physiological groups of bacteria and archaea that utilize sulfate as the final electron acceptor in anaerobic respiration are known as sulfate-reducing prokaryotes. Three archaeal species are known to use sulfate respiration and all are members of the genus *Archaeoglobus*. They have received relatively little attention with respect to metal interactions. Thus, our review will rely heavily on the metal interactions with Gram-positive and Gram-negative sulfate-reducing (SRB) eubacteria.

The SRB are chemolithotrophs found in anaerobic environments containing compounds such as toxic metals and metalloids. Resistance to redox-active metals by SRBs is attributed to reduction processes that produce metal ions of decreased solubility and decreased toxicity. Divalent cationic transition and heavy metals in the environment are precipitated by sulfide produced as a product of dissimilatory sulfate reduction.

Although a paper by Woolfolk and Whiteley [102] reported the capability of metal reduction by sulfate-reducing bacteria, this activity was not addressed until bioremediation using anaerobic bacteria was pursued. In the last several decades many papers have been published concerning metal reduction by sulfate reducers. These activities have been summarized in numerous reviews (e.g., [8–10, 16, 50, 56]).

The group of sulfate-reducing microbes is highly diverse. Fifty nine genera and 220 species are currently recognized [33]. Of the latter, fifty seven species belong to the genus *Desulfovibrio*.

(http://www.bacterio.net/).

This review examines the diversity of systems that are employed by SRB to interact with metals and metalloids in the environment. Metal stress response in SRB is a complex process. The review examines some of the major mechanisms enabling this group of anaerobes to inhabit diverse environments containing toxic metals. We make particular note of the use of SRB to remediate metal contaminated sites. We also comment of the interest to produce metals for commercial applications.

4.2 Mechanisms of Metal Resistance

4.2.1 Reactions Mediated by Biogenic Sulfide

Sulfidogenic activities in nature are attributed to abiogenic geochemical processes or to biological transformations. Due to the quantity of sulfide produced by dissimilatory sulfate reduction, the physiological group of sulfate reducers is responsible for most of the biosulfide produced. Since there are only three genera of Archaea and 56 genera of Bacteria, biogenic sulfide production is essentially a bacterial activity [6]. Biogenic hydrogen sulfide is a potent reducing agent whose formation accounts for the reduction of metals and metalloids. It forms highly insoluble metal sulfides, as well [11, 36, 99].

4.2.2 Respiratory Metal Reduction

While the hallmark electron donors for sulfate reducing SRB include pyruvate, lactate, H_2, and formate, specific strains are capable of using over 75 different substances to support growth [32]. These substances are transformed by the SRB using a vast array of electron transport components. These include dehydrogenases, reductases, cytochromes, ferredoxins, and flavodoxins [6, 10, 33, 74]. These enzymatic capabilities enable the SRB to grow in environments where toxic levels of metals or metalloids exist.

The metal reductases in SRB lack metal or metalloid specificity. This is observed from the activities of isolated hydrogenases and cytochromes listed in Table 4.1. This non-specific enzymatic reduction of metal ions is not limited to SRB. It has also been reported for hydrogenase, cytochrome, nitrate reductase and catalase isolated from other anaerobic and facultative-anaerobic bacteria [7, 8].

The reduction of redox-active metal cations by SRB often results in the production of metallic nanoparticles consisting of the elemental form. A selection of these metal (loid) nanoparticles produced by SRB is listed in Table 4.2. This production of nanoparticles is not unique to SRB. Many taxonomically diverse bacteria produce metallic nanoparticles either in the cytoplasm, periplasm, or extracellular region [7].

4.3 Reduction of Metalloids

4.3.1 Selenate and Selenite Reduction

Selenate shares close chemical similarities with sulfate. Thus, it can replace sulfate in its reaction with ATP sulfurylase, the first enzyme in the sulfate reduction pathway [79]. The resulting adduct, adenosine 5′-selenophosphate (APSe), is then reduced to

Table 4.1 Metal(loid) reductions attributed to hydrogenases and cytochromes from sulfate-reducing bacteria

Protein	Bacteria	Element reduced	Reference
Cytochromes			
c_3	*D. fructosivorans*	Tc(VII)	[25]
c_3	*D. gigas*	Fe(III)	[55]
c_3	*Dsm. norvegicum*	Cr(VI)	[64]
c_3	*Dsm. norvegicum*	Fe(III)	[55]
c_3	*D. vulgaris* Hildenborough	Fe(III)	[57]
c_3	*D. vulgaris* Hildenborough	Se(VI)	[1]
c_3	*D. vulgaris* Hildenborough	U(VI)	[57]
c_7	*Dsf. acetoxidans*[a]	Cr(VI)	[55]
c_7	*Dsf. acetoxidans*	Fe(III)	[55]
c_7	*Dsf. acetoxidans*	Mn(IV)	[55]
Hydrogenase			
	D. deslfuricans G 20[b]	Pd(II)	[51]
	D. deslfuricans G 20	Tc(III)	[53]
[Fe] hydrogenase	*D. fructosivorans*	Tc(VII)	[25]
[Fe] hydrogenase	*D. vulgaris* Hildenborough	Cr(VI)	[64]
[NiFe] hydrogenase	*D. fructosivorans*	Cr(VI)	[20]
[NiFeSe] hydrogenase	*Dsm. norvegicum*	Cr(VI)	[64]

[a]A sulfur-reducing bacterium
[b]Reclassified recently as *D. alaskensis*

Table 4.2 Production of metal(loid) nanoparticles by *D. desulfuricans*

Nanoparticle produced	Metal/metalloid reduced	Reference
Se^0	SeO_4^{2-}	[94]
	SeO_3^{2-}	[94]
Re^0	ReO_4^-	[105]
Au^0	$HAuCl_4$	[26]
Pd^0	$Pd(NH_3)_4Cl$	[51]
Pt^0	PtO_2	[81]
Rh^0	Rh^{3+}	[106]
UO_2	UO_4^{2+}	[105]

selenite [27]. This is the same reaction catalyzed in the activation of sulfate to APS and subsequent reduction to sulfide. However, while sulfite is reduced to hydrogen sulfide, selenite is reduced to elemental selenium rather than hydrogen selenide.

SRB cannot conserve energy and grow from the reduction of selenate to selenite or from selenite to elemental selenium. They do not have a strict need for the elemental form to synthesize organic compounds essential for growth. The process cannot be considered as detoxification, either. The reduction of selenite with thiol groups results in toxicity, not only by reacting with thiol groups in general [71], but by generating superoxide and other reactive oxygen species as well [86].

Despite the propensity of selenite to react with thiol groups and other redox sensitive substances, e.g., vitamin C, selenate and selenite reduction does not occur indiscriminately in actively growing cultures. Evidence suggests that the ability to reduce selenate and selenite in large quantities to Se(0) is inducible and regulated by the cells. This phenomenon was studied in *D. desulfuricans* grown in a semi defined formate/fumarate and sulfate-free medium with cysteine (0.1 mM) added as the sole sulfur source [94].

Selenate and selenite concentrations of a few micromolar inhibit the growth of SRB. But cultures can be adapted to grow in the presence of significantly higher concentrations. For example, Tomei et al. [94] adapted *D. desulfuricans* to grow in the presence of 10 mM selenate or 0.1 mM selenite. These cells grew exponentially in the presence of the latter concentrations. The only visible effect was the loss of cell shape, i.e., the curved rods became straight, when grown in the presence of selenite, but not in the presence of selenate.

Adapted *D. desulfuricans* cultures reduced both selenate and selenite to elemental selenium, but only after exponential growth ceased and the cells entered into early stationary growth phase. The elemental selenium accumulated inside the bacterial cells, see Figs. 4.1, 4.2 and 4.3. Interestingly enough, evaluation of thin sections of bacterial cells growing in the presence of selenate indicated large deposits of Se(0) in the region of the periplasm, while the cells grown in the presence of selenite accumulated the elemental selenium in the cytoplasm.

D. desulfuricans actively growing on sulfate as the terminal electron acceptor forms particles in the presence of selenite. These particles contain S as well as Se [104]. This phenomenon can be explained by the abiotic reaction between the sulfide produced by the bacteria and the selenite present in the growth medium.

$$2H_2S + SeO_3^{2-} + 2H^+ \rightarrow SeS_2 + 3H_2O$$

4.3.1.1 Practical Applications

The observation that *D. desulfuricans* removes selenium from solution as it reaches stationary phase can be used as a strategy for bioremediation of waters contaminated with selenate and selenite [4]. From a biotechnology perspective, immobilized cells of *D. desulfuricans* in column tests readily remove selenate as Se(0) [96]. The presence of Se(0) at the interior of the polyacrylamide gel may suggest involvement of cell respiration because cells remain metabolically active and multiply when placed in polyacrylamide gels. With *D. desulfuricans*, a sulfide rich

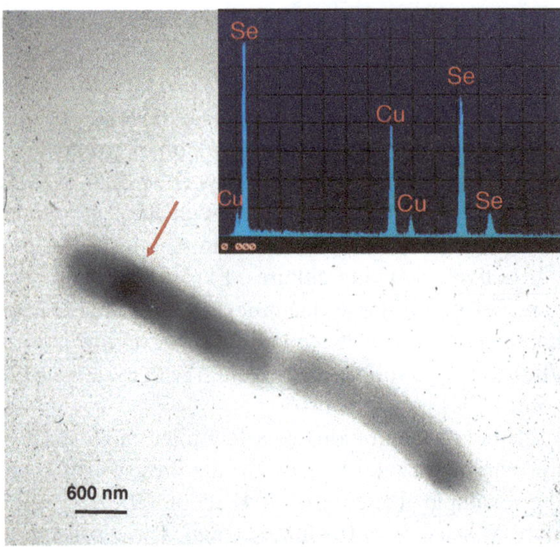

Fig. 4.1 *D. desulfuricans* was grown in a formate/fumarate medium with 0.1 mM sodium selenite. Cells were removed from the culture in early log phase and examined under a JEOL 2000EX scanning-transmission electron microscope as unstained whole mounts. The dark internal structure at the arrow was determined to be selenium by energy dispersive X-ray (EDX) microanalysis using a Tracor Northern 550 spectrometer. The use of copper grids accounts for the copper peaks. Methods for cultivation and electron microscopy are described by Tomei et al. [94]

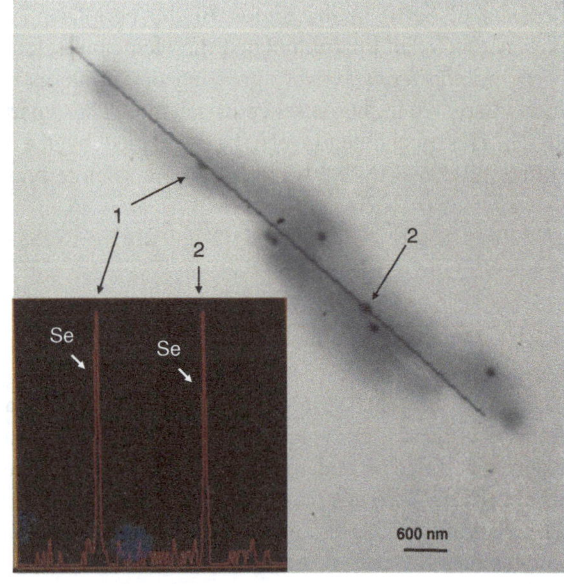

Fig. 4.2 *D. desulfuricans* was grown in a formate/fumarate medium with 0.01 mM sodium selenate. The *small dark structures* inside the cells were determined to be selenium. Scanning along the line revealed Se in the granules identified as "*1*" and "2". Methods for cultivation and electron microscopy are described by Tomei et al. [94]

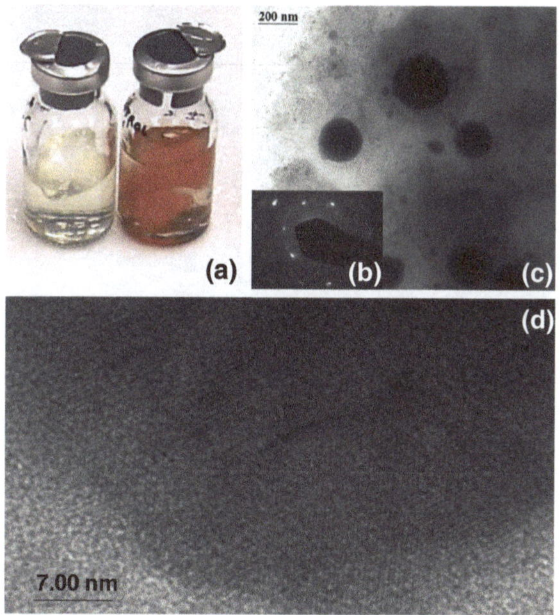

Fig. 4.3 *D. desulfuricans* was grown for 2 days in a lactate/sulfate medium containing 0.1 mM sodium selenate. *Inset* (*A*) shows a bottle of uninoculated culture medium with no deposits and the bottle inoculated with the bacterium shows red deposits in the culture after 2 days. The *dark spheres* inside the cells are Se-rich (Se/S) solid solution crystals. *Insert* (*B*) is a selected-area electron diffraction (SAED) pattern indicating crystalline character of the Se deposits (*C*). High-resolution image from a nanocrystal of Se (monoclinic) (*D*). Methods for the cultivation and electron microscopy are described by Tucker et al. [96], Xu and Barton [104], respectively

environment enhances the conversion of selenite to Se(0) [52]. Elevated concentrations of sulfide disperse colloidal Se particles and a method for quantitation of Se (0) production has been published [13].

Complex reactions occur with selenate and selenite in SRB. Additional research is required to understand the mechanism of selenium-SRB interactions.

4.3.2 Tellurate Reduction

Resistance to tellurite is found in numerous bacterial species. In some instances, bacteria detoxify tellurite by glutathione, methyltransferase reactions or by the TeR plasmid-mediated process [90]. Tellurium (Te), like Se, has stable oxidation states of +VI (tellurate), +IV (tellurite), 0(tellurium) and −II (telluride), however, relatively few microorganisms have been reported to couple anaerobic tellurium respiration to growth. Thus far, the only sulfate-reducing bacterium that enzymatically reduces tellurite is *D. desulfuricans* [52]. The reduction of Te(VI) by *D. desulfuricans* is shown in the image from electron microscopy in Fig. 4.4.

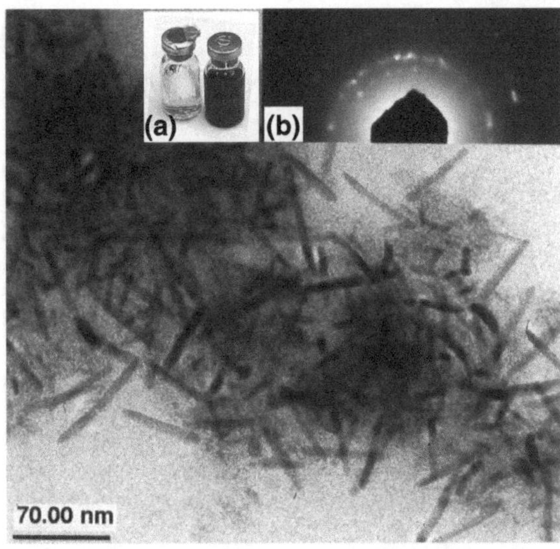

Fig. 4.4 TEM image of tellurite (TeO2) nanocrystals produced by *D. desulfuricans* grown for 2 days in a lactate/sulfate medium containing 1 mM tellurate. *Inset A* is a clear bottle of uninoculated medium and the *dark bottle* shows the black insoluble tellurite crystals. *Inset B* is a selected-area electron diffraction (SAED) pattern indicating the crystalline order of tellurite. Methods for cultivation and electron microscopy are described by Tucker et al. [96], Xu and Barton [104], respectively

4.3.3 Arsenate Reduction

While numerous bacteria reduce arsenate [88], there are relatively few reports for arsenate reduction by SRB. Sulfate reducers that also reduce arsenate include *Desulfosporosinus auripigmenti* [69]; formerly *Desulfotomaculum auripigmentum*; [87], *Desulfomicrobium* Ben-RB, *Desulfovibrio* Ben-RA [62], *D. desulfuricans* (reclassified as *D. alaskensis*) G20 [43]. While *D. alaskensis* G20 and *Desulfovibrio* strain Ben-RA reduce arsenate only when sulfate is supplied, *Desulfomicrobium* Ben-RBcan grow with arsenate as the sole electron acceptor. In *Desulfosporosinus auripigmenti*, arsenate respiration is independent from sulfate reduction and arsenate reduction precedes reduction of sulfate. Electron micrographs indicating reduction of arsenate by *D. desulfuricans* is given in Fig. 4.5.

For bacterial reduction of arsenate, there appear to be two approaches. One is the detoxifying *ars* system and the second is a respiring membrane-associated reduction. Reduction of arsenate in *D. alaskensis* G20 is attributed to *arsC* gene and the *arsRBCC* that are located at different sites in the genome [43]. Putative genes, with some limitations, for the *ars* system are also present in *D. vulgaris* Hildenborough, *Desulfotalea psychrophila,* and *Archaeoglobus fulgidis* [16, 43]. Genes of the *ars*

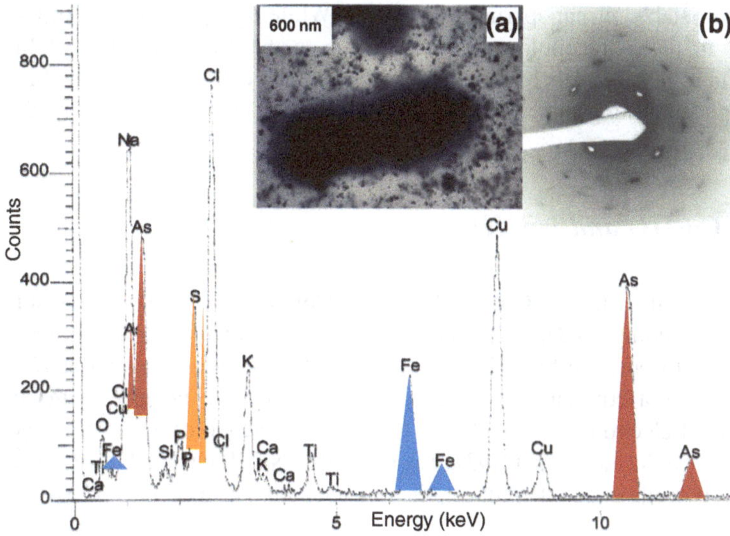

Fig. 4.5 EDX microanalysis of extracellular arsenic precipitates produced by *D. desulfuricans* grown for 2 days in a lactate/sulfate medium containing 1 mM sodium arsenate as seen in *inset* (*A*). Peaks indicating presence of arsenic are highlighted in *red*, iron in *blue* and sulfur in *orange*. *Inset* (*B*) is a selected-area electron diffraction (SAED) pattern indicating crystalline order of arsenic deposits. Methods for cultivation and electron microscopy are described by Xu and Barton [104], Tucker et al. [96], respectively

operon are also found in *Desulfuromonas acetoxidans*, a sulfur reducing bacterium [43]. Details of the respiratory-coupled arsenate reduction that supports growth remain to be established.

4.4 Reactions with Transition Metals

4.4.1 Au(III) and Au(I)

In an initial report, gold as Au(III) was reported to be precipitated by *D. desulfuricans* with the production of Au(0) in the extracellular region [23]. A subsequent paper indicated that *D. desulfuricans* will reduce $HAuCl_4$ at pH 7 to crystalline Au(0). These Au nanoparticles are found in the periplasm, on the cell surface and in the cytoplasm [26]. This reduction by *D. desulfuricans* is coupled to the oxidation of H_2.

A sulfate-reducing bacterium from a gold mine in South Africa has been found to produce elemental gold [Au(0)] from gold-thiosulfate [49]. This process is interesting because it is proposed that hydrogen sulfide produced by the sulfate reducing bacteria destabilizes the $Au(S_2O_3)_2^-$ complex resulting in the reduction of

Au(I) to octahedral nanoparticles of Au(0). While nanoparticles of Au(0) may be formed in the extracellular surrounding due to the reducing effect of biosulfide production, there were <10 nm nanoparticles on the cell surface and within the cell envelope suggesting that electron transport activity may also be involved.

4.4.2 Co(III) and Ni(III)

D. vulgaris interacts with Co(III) chelated as $CoEDTA^{-}$. Cobalt is reduced to Co(II) and remains complexed as $CoEDTA^{2-}$ [15]. The primary mechanism for reduction of Co(III) is attributed to biosulfide production where CoS precipitates as *D. vulgaris* uses sulfate as a terminal electron acceptor. *D. vulgaris* does not grow with $CoEDTA^{-}$ as the final electron acceptor. Since Ni(III) in the active center of hydrogenase of *D. gigas* is reduced to Ni(II) [93], it is likely that Ni(III) chelated to EDTA could be reduced by cells of sulfate reducers in a manner similar to chelated Co(III).

4.4.3 Cr(VI)

The reduction of highly soluble and chemically reactive Cr(VI) to less reactive Cr (III) has been demonstrated with metabolically active cells of sulfate reducers as well as with cell-free extracts. Using a consortium of bacteria isolated from metal-refinishing waste water, Fude et al. [35] demonstrated that Cr(IV) reduction occurred only with the addition of sulfate. The formation of Cr(III) was attributed to sulfide production.

Biofilms containing sulfate-reducing bacteria have also been observed to reduce Cr(IV) as a result of sulfidogenic metabolism [85]. An isolate, *Desulfovibrio* sp. Oz7, was reported to reduce Cr(VI) to Cr(III) with lactate or H_2 as the electron donor, using a bicarbonate buffer [52]. The resistance of *D. desulfuricans* G20 (recently reclassified as *D. alaskensis* G20) has been demonstrated to involve thioredoxin. Reduction of Cr(VI) by thioredoxin may alleviate chromate toxicity in the cytoplasm [44].

Washed or immobilized cells have been useful to avoid the chemical reaction of hydrogen sulfide with chromate. Cr(VI) reduction by *D. desulfuricans* immobilized in polyacrylamide gel revealed that cells continuously exposed to 0.5 mM Cr(VI) remained metabolically active. Cr(III) was precipitated on the surface and inside of the gel [96]. An image of Cr(VI) reduced by D. *desulfuricans* is given in Fig. 4.6.

When cells of *D. vulgaris* were palladized, they were found to readily reduce Cr (VI), chromate, to Cr(III) at pH 3 [40]. A process of commercial potential was reported by Humphries et al. [41] using *D. vulgaris* NCIMB 8303 immobilized in agar. From a survey of various sulfate-reducing bacteria, Michel et al. [64] found that the highest rates of chromate reduction occurred with *Desulfomicrobium norvegicum* DSM 1741.

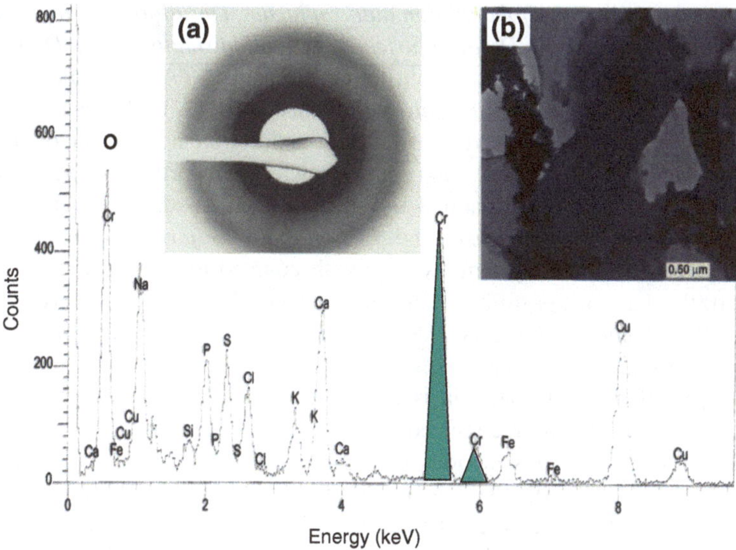

Fig. 4.6 EDX microanalysis of chromium precipitates associated with cells of *D. desulfuricans* grown for 2 days in a lactate/sulfate medium containing 0.1 mM potassium chromate. Cr peaks are highlighted in *green*. *Inset* (*A*) is a selected-area electron diffraction (SAED) pattern indicating the lack of crystalline order of chromium. *Insert* (*B*) is an electron diffraction pattern from the Cr precipitates indicating lack of crystalline pattern. Methods for electron microscopy are described by Xu and Barton [104]

Cell-free experiments have provided information on two proteins that show chromate reductase activity. Chardin et al. [20] reported that [Fe], [NiFe], and [NiFeSe] hydrogenases isolated from species of *Desulfovibrio* or *Desulfotomaculum* functioned as chromium reductases. The [3Fe-4S] ferredoxin from *D. gigas* was used as a model to address chromium reduction. They found that the low redox [Fe-S] clusters account for Cr(VI) reduction. This is proposed to be the mechanism for reduction by hydrogenases [20].

Cytochrome c_3 isolated from *D. vulgaris* Hildenborough (ATCC 29579) reduced chromate [60]. The cytochrome c_7 from *Desulfuromonas acetoxidans* has provided interesting information about chromium reduction [2]. Cytochrome c_7 has three heme groups and Cr(VI) binds onto the protein at a site that is closer to heme IV than to heme III or heme I. This specific binding site would support the proposal that cytochrome c_7 is a chromium reductase.

4.4.4 Fe(0) and Fe(III)

Dissimilatory iron reduction is associated with a physiological group of SRB. While several bacteria can reduce Fe(III) to Fe(II), not all of them can couple iron reduction to growth. *Desulfobacterium autotrophicum*, *Desulfobulbus propionicus*, *D. baculatus*, *D. baarsii*, *D. desulfuricans*, *D. sulfodismutans* and *D. vulgaris* reduce Fe(III) when chelated with nitrilotriacetic acid (NTA) or as Fe(III) oxide [57]. However, *Desulfobacter postgatei* will reduce Fe(III)-NTA but not iron oxide. The sulfate reducers that reduced Fe-NTA or Fe-oxide were not capable of coupling this electron flow to Fe(III) and cell growth. Similarly, *D. frigidus* and *D. ferrireducens* displayed capability of reducing Fe(III) without growth [100].

Desulfuromonas svalbardensis and *Desulfuromonas ferrireducens* have dissimilatory iron reduction systems with growth coupled to acetate as the electron donor [100]. Many of the sulfate reducers are incapable of reducing Fe(III). These include *Desulfobacter curvatus*, *Desulfomonile tiedjei* and *Desulfotomaculum acetoxidans* [57].

Corrosion of metal by bacteria is frequently referred to as microbial induced corrosion (MIC). SRB are recognized to play a primary role in the anaerobic iron metal corrosion. Numerous theories on the mechanisms of this MIC have been developed [12, 31, 68, 70]. Using an environmental scanning electron microscope, the differential biofilm structures of *D. vulgaris* in corrosion of ferrous, carbon steel and steel wire were evaluated with respect to the cathodic depolarization theory [37]. Recently, a new mechanism for ferrous corrosion has been proposed by Enning et al. [31]. This involves electroconductive ferrous sulfide crusts where electrons are released from the formation of the mineral pyrite from FeS and hydrogen sulfide.

IronIron corrodes in the absence of oxygen to produce the ferrous cation and the concomitant formation of hydrogen gas:

$$Fe^0 + 2H^+ \rightarrow Fe^{2+} + H_2$$

The liberated hydrogen gas is a natural growth substrate for a wide variety of bacteria capable of coupling its oxidation to energy conservation and growth. In the case of the sulfate-reducing bacteria, hydrogen consumption is coupled to the reduction of sulfate to sulfide. The removal of the H_2 by bacteria is historically known as biologically induced corrosion by cathodic depolarization. The question then becomes whether corrosion is the result of bacterial production of sulfide or vice versa. The issue was addressed in part by Tomei and Mitchell [95]. The authors separated the issue of hydrogen consumption from that of sulfide production by growing *Desulfovibrio* in a sulfate-free medium with fumarate as the terminal electron acceptor. Their studies revealed that: (1) The bacteria readily grew on the H_2 coming off the metal as the sole source of energy. (2) The corrosion rates were faster in the presence of the bacterium that in sterile growth medium. (3) The corrosion rates in the presence of bacteria were a function of temperature, i.e., maximum corrosion rates occurred at the optimum growth temperature of the bacterium and decreased at higher temperatures. (5) However, the highest observed corrosion rates were significantly lower than those observed under environmental conditions.

4.4.5 Hg(II)

Mercury is a toxic metal. A common strategy for bacteria is to employ the mer operon as one of the defense mechanisms. This operon encodes proteins for the detection, binding, transport and reduction of mercury. The reduction of Hg(II) to Hg(0) is attributed to an NADPH-dependent enzyme. As indicated in a review by Bruschi et al. [16], putative genes for mercury reduction have been found in *D. vulgaris* Hildenborough, *D. alaskensis* G20, *Desulfotalea psychrophila,* and *Archaeoglobus fulgidis*. However, direct experimentation of the *mer* operon functioning in SRB has not been demonstrated.

Another detoxifying system involves mercury methylation and it is of considerable importance because it contributes to movement of mercury from the environment and as a result contributes to reduced human toxicity [80]. The initial reports on mercury methylation by *D. desulfuricans* LS indicated the involvement of the acetyl coenzyme A pathway [22] while Ekstrom et al. [29] have demonstrated that sulfate-reducing bacteria that lack the acetyl-CoA pathway are capable of mercury methylation.

To alleviate the problems of sulfide precipitation of mercury as HgS, washed cells are frequently used to follow the mercury methylation reactions localized in the cytoplasm. Gilmour et al. [38] indicate that over 50 % of the *Desulfovibrio* strains tested have the capability of producing methylmercury. From a study involving 59 species of *Desulfovibrio*, Graham et al. [39] determined that mercury methylation is species-specific and is not restricted to a given phylogeny. The 11 species of *Desulfovibrio* demonstrated to methylate mercury are *D. aespoeensis* DSM 100631, *D. aftricans* DSM 2603, *D. alcoholivorans* DSM 5433, *D. alkalitolerans* DSM 16529, *D. carbinoliphilus* DSM 17524, *D. desulfuricans* ND 132, *D. desulfuricans* DSM 642, *D. piger* DSM 749, *D. psychrotolerans* DSM 19430, *D. sulfodismutans* DSM 3696, and *D. tunisiensis* DSM 19275.

Recently, it was shown that for the mercury methylation process in *D. desulfuricans* ND132, genes *hgcA* and *hgcB* are required [72]. These genes encode a corrinoid protein, HgcA protein, HgcB protein, and a 2[4Fe-4S] ferredoxin that function as a methyl carrier and an electron carrier for corrinoid protein reduction, respectively. It appears that sulfate reducing bacteria and archaea that methylate mercury also have these genes.

4.4.6 Mo(VI)

Molybdate, Mo(VI), is reduced by washed cells of *D. desulfuricans* DSM 642 to Mo(IV). MoS_2 is formed in the presence of sulfide [97]. The black precipitate of MoS_2 was deposited extracellularly and was confirmed to be molybdenite (MoS_2) by electron diffraction.

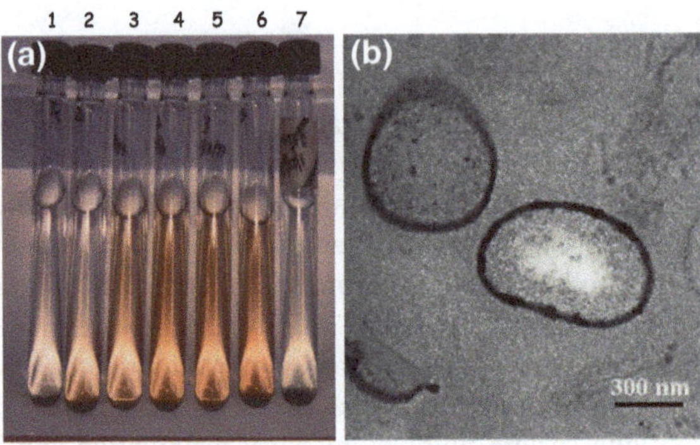

Fig. 4.7 The *red-brown color* indicates reduction of molybdate by *D. desulfuricans* grown for 2 days in a lactate/sulfate medium containing sodium molybdate. The *black precipitate* at the *bottom* of the culture tubes contains MoS2 and FeS, (*A*). Growth in each tube was at a different concentration of sodium molybdate. Amount of sodium molybdate added to each tube: tube 1, none; tube 2, 0.05 mM; tube 3, 0.1 mM; tube 4, 0.5 mM; tube 5, 1.0 mM; tube 6, 1.5 mM; and tube 2.0 mM. Thin section of cells of *D. desulfuricans* growing in sodium molybdate showing the localization of MoS2 in the periplasm. Methods for cultivation and electron microscopy are described by Biswas et al. [14], Xu and Barton [104]

Enzymatic reduction of Mo(VI) was also accomplished using *D. desulfuricans* immobilized in polyacrylamide gel [96]. From electron microscopy of thin sections of cells, the reduction of Mo(VI) with formation of MoS_2 was found to occur in the periplasm of *D. desulfuricans* and *D. gigas* [14]; see Fig. 4.7. Additionally, MoS_2 was intermittently distributed on the surface of the cells of *D. gigas*. Cells of *D. desulfuricans* have been reported to produce MoS_2 from elemental molybdenum powder [21].

4.4.7 Mn(IV)

Dissimilatory reduction of manganese has been reported for both Gram-positive and Gram-negative sulfate reducing bacteria. Growth has been coupled to the reduction of Mn(IV) by *Desulfotomaculum reducens* Strain MI-1 [91]. *Desulfotomaculum geothermicum*, a spore forming bacterium, reduces manganese oxide [83]. The dissimilatory reduction of manganese oxide [MnO_2; (Mn(VI)] to $MnCO_3$ [Mn(II)] is coupled to lactate oxidation, which supports growth of *Desulfovibrio* strain $CY1^T$. The latter is an isolate from sediments of waters streaming from lead, zinc and antimony mines [84].

Cytochrome c_7 isolated from a sulfur-reducing bacterium that produces sulfide, *Desulfuromonas acetoxidans,* readily reduces manganese (IV) oxide to Mn(II) [55]. Since cells of *D. desulfuricans* and *Desulfomicrobium baculatum* (*D. baculatus*) reduce Mn(IV) [58], it is likely that cytochrome c_3 from these organisms will also reduce manganese, as these bacteria have cytochromes with bishistidinyl heme iron coordination.

In nature, sulfide produced by sulfate-reducing bacteria reacts with manganese oxide (MnO_2) resulting in the precipitation of MnS and production of S(0). This dependency on sulfate reducers for the reduction of manganese has been reported by Burdige and Nealson [17]. The process is proposed to be important in naturally-stratified environments such as the Black Sea.

4.4.8 Pb(II)

The removal of cationic lead from industrial wastewater has been demonstrated using biogenic sulfide produced by sulfate-reducing bacteria [92]. Removal of 99 % of the soluble lead (20 mg/L) occurred if the sulfide:lead molar ratio was 3:1 and the pH was 8.0. The biosorption of lead by the biomass of sulfate reducing bacteria has been reported by El Bayoumy et al. [30]. The optimal ratio for lead removal to the amount of sulfate reducing bacterial biomass is 0.3. This biosorption of lead would include the precipitation of lead sulfide from biogenic sulfide as well as binding to cells. While living cells may not have much lead binding, dead cells included in the biomass could provide for significant biosorption of cationic lead.

4.4.9 Pd(II)

D. desulfuricans ATCC 29577 reduced 0.5 mM Pd(II) to Pd(0) when $Pd(NH_3)_4Cl$ was added to cells energized with pyruvate, formate or H_2 as electron donors [51]. The observation of thin sections of cells reducing Pd(II) by transmission electron microscopy revealed the localization of Pd(0) deposits in the periplasm. The presence of Pd(0) was established by EDS and X-ray analysis. The involvement of periplasmic hydrogenase was implicated since H_2 driven reduction of Pd(II) could be inhibited by 0.5 mM Cu(II).

In a related study, cells of *D. desulfuricans* NCIMB 8307 were immobilized on a Pd-Ag membrane [106]. The membrane was perfused with a solution composed of 2 mM $PdCl_4^{2-}$ and 10 mM HNO_3 (pH 2) and hydrogen was supplied by an electrochemical process. Pd(0) produced by this electrobioreactor was established by polography. Additionally, formation of Pd(0) was demonstrated with cells of *D. desulfuricans* NCIMB 8307 suspended in 10 mM HNO_3 containing 2 mM $PdCl_4^{2-}$ with H_2 as the electron donor.

The mechanism of Pd(0) production from Pd(II) in *Desulfovibrio* appears to involve hydrogenase. The reduction of Pd(II) by *D. fructosivorans* was examined using hydrogenase-deficient strains. As reviewed by Mikheenko et al. [65], *D. fructosivorans* has two periplasmic hydrogenase (a NiFe hydrogenase and a Fe hydrogenase), a cytoplasmic NADP-reducing hydrogenase and a NiFe hydrogenase localized in the cytoplasmic membrane. Using hydrogenase-negative mutants of *D. fructosivorans*, Mikheenko et al. [65] was able to demonstrate that the reduction of Pd(II) can be accomplished by hydrogenase enzymes localized in the periplasm or in the cytoplasmic membrane. Under the acidic conditions (pH 2) of the reduction process, *c*-type cytochrome in the periplasm of *D. fructosivorans* was unable to reduce Pd(II) to Pd(0).

4.4.10 Pt(IV) and Rh (III)

Using living cells of a sulfate-reducing bacterium, Rashamuse and Whiteley [81] demonstrated that the reduction of Pt(IV) was dependent on hydrogenase activity and Pt was precipitated in the periplasm. The reduction of Pt(IV) and Rh(III) by *D. desulfuricans* NCIMB 8307 has been reported by Yong et al. [106]. Cells of *D. desulfuricans* in an electrobioreactor were exposed to an industrial precious metal processing stream containing <5 mM Pt(IV), Rh(III), and Pd(II). With hydrogen generated electronically and a residence time of 10 to 20 min at pH 2.5, the removal of Pt(IV), Rh(III), and Pd(II) was 99, 75 and 88 %, respectively.

4.4.11 Re(VII)

Rhenium as Re(VII) is frequently used in laboratory studies because it is a chemical analogue of Tc(VII), a fission product of uranium. Cells of *D. desulfuricans* strain 642 readily reduce the oxyanion ReO_4^- with the production of Re(0) at the surface of the cell [105]. The rate of Re(VII) reduction with H_2 as the electron donor is greater than the reduction rate with lactate.

4.4.12 Tc(VII)

Technetium is a fission product of ^{235}U. It exists as the highly stable pertechnetate ion (TcO_4^-) as a product of the nuclear fuel cycle. Not only is the pertechnetate ion highly mobile, it readily enters the food chain by rapid assimilation into plants by the sulfate uptake system [19]. To detoxify the environment, several bacteria were found to reduce Tc(VII) to insoluble black oxide of TcO_2, Tc(IV) [50]. Resting cells of *D. desulfuricans* ATCC 29577 were found to reduce ammonium pertechnetate (NH_4TcO_4) with precipitation of insoluble Tc in the periplasm [53]. Lloyd et al. [53] were able to determine that the insoluble precipitate was primarily Tc without sulfur. This mechanism was distinct from an earlier report where insoluble sulfides of Tc(VII) and Tc(IV) were reported [75]. Since 0.5 mM Cu(II) inhibited the reduction of Tc(VII) with H_2 as the electron donor, it was inferred that deposition of Tc by *D. desulfuricans* involved periplasmic hydrogenase [54].

To exclude possible reduction of Tc(VII) by sulfide produced by sulfate reducers, *Desulfovibrio fructosivorans* DSM 3604 was grown in a medium with fructose as the electron donor and fumarate as the electron acceptor. These resting cells were incubated with 1 mM Tc(VII). High levels of Tc(VII) reduction occurred with H_2 as the electron donor and only minimal reduction occurred with fructose, lactate, pyruvate, or formate [25]. The resting cells with H_2 as the electron donor followed Michaelis–Menten kinetics and had an apparent K_m of 2 mM for Tc(VII) and a maximal velocity of 7 mmoles of Tc(VII)/g dry wt/h of incubation. In comparison, the apparent K_m reported for cells of *D. desulfuricans* was 0.5 mM [54].

D. desulfuricans has been reported to reduce Tc(IV) to the base metal Tc(0). This reduction involves hydrogenase activity [52]. In cell-free studies, De Luca et al. [25] report that the Ni-Fe hydrogenase isolated from the periplasm of *D. fructosivorans* was the reductase that converted Tc(VII) to soluble Tc(V) or insoluble Tc(IV). Cytochrome c_3 purified from *D. fructosivorans* was unable to reduce Tc(VII) but when purified Ni-Fe hydrogenase was added to the reaction the rate of reduction was greater than when Ni-Fe hydrogenase were used alone. The mechanism involving cytochrome c_3 in this reduction is unresolved.

4.4.13 V(V)

The enzymatic reduction of pentavalent vanadium (VO_3^-, vanadate) to trivalent vanadium (V^{3+}) is attributed to only a few bacteria [61]. Bacteria with respiratory coupled vanadium reduction include *Clostridium pasteuranium*, *Pseudomonas vanadiumreductans*, *Pseudomonas issachenkovii*, *Shewanella onedensis*, MR-1 and *Geobacter metallireducens* [28].

There is the singular report by Wolfolk and Whiteley [102] that *Desulfovibrio fructosivorans* coupled the reduction of vanadate to H_2 oxidation with 0.57 mmoles of H_2 consumed/min/mg protein of bacteria. Cultures of anaerobes that would reduce orthovanadate or metavanadate have been found associated with tubeworms from hydrothermal vent fields in the eastern Pacific Ocean [24].

4.4.14 Zn(II)

Zinc in the form of Zn^{2+} is a trace nutrient for growth but at elevated concentrations is toxic. Biogenic hydrogen sulfide produced by sulfate-reducing bacteria reacts with Zn^{2+} resulting in the precipitation as ZnS. Natural biofilms containing sulfate-reducing bacteria are known to form ZnS (sphalerite) [47] In several instances mixed cultures of bacteria resistant to heavy metals have been shown to precipitate zinc as a sulfide because sulfate-reducing bacteria are present [3, 76]. In The Netherlands, ground water from the Budelco zinc refinery is treated by sulfide precipitation. As reviewed by Hocking and Gadd [42], this commercial process uses

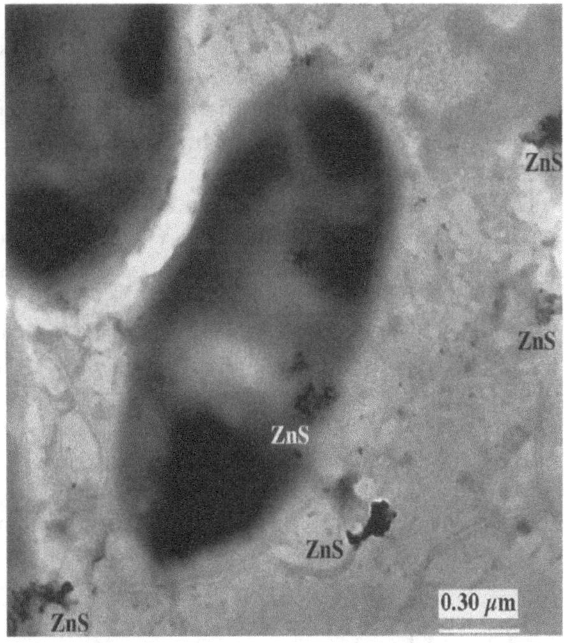

Fig. 4.8 TEM image of ZnS (sphalerite) nanocrystals and two cells *D. desulfuricans*. Cells were grown for 2 days in a lactate/sulfate medium containing 0.1 mM znic sulfate. Methods for cultivation and electron microscopy are described by Tucker et al. [96], Xu and Barton [104], respectively

THIOPAQ® technology [http://en.paques.nl/products/featured/thiopaq] where a two-phase bioprocess is used. In the first phase, sulfate-reducing bacteria produce 2. 5 tons of H_2S each day with the precipitation of ZnS and in the second stage, aerobic bacteria are used to oxidize the sulfide to elemental sulfur [S(0)].

D. desulfuricans cultivated in Postgate medium with 0.5 % tryptone and 0.4 % yeast extract, formed ZnS on the cells and in the extracellular medium (Fig. 4.8). Evaluation of the ZnS deposits revealed a crystalline form (Fig. 4.9). Since ZnS was not seen in inorganic salts medium of LeGall, it appears that the level of Zn(II) in the tryptone and yeast extract contributed to the observed precipitation of ZnS.

4.5 Reactions with Actinides

4.5.1 Pu(VI) and Pu(IV)

D. äspöensis DSM 10613 isolated from the granitic rock aquifer was effective in binding Pu(VI) and Pu(IV) polymers [67]. Prior to the production of Pu(V) from Pu (VII), Pu binds onto the cell. The surface of the cell binds Pu(V) only weakly and Pu(V) readily dissociates into the medium surrounding the cell. However, it appears that some plutonium is found in the cytoplasm of the cell. Evaluation of these reduction reactions is difficult because the speciation of Pu is markedly complex.

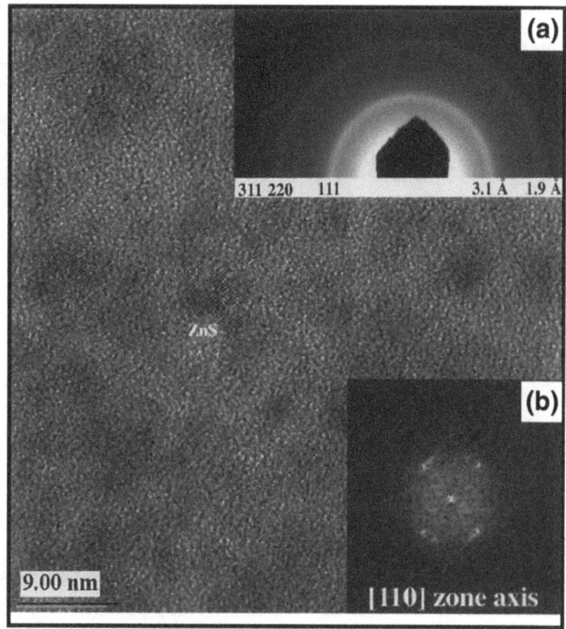

Fig. 4.9 Evaluation of extracellular ZnS nanocrystals shown in Fig 4.8. Image is a high resolution TEM of the aggregates of sphalerite showing lattice fringes within the nanocrystals. Insert (**a**) is an electron diffraction pattern from the ZnS aggregates. Insert (**b**) is the Fourier Transform of a nanocrystal (labeled ZnS) along zone axis [110]. Methods for electron microscopy are described by Xu and Barton [104]

4.5.2 U(Vi)

As a consequence of the development of the nuclear industry, uranium in the environment is of concern because it remains at the sites where uranium ore was processed. The role of bacteria and especially sulfate reducing bacteria has been discussed in recent reviews [5, 10, 101]. Application of anaerobic bacteria to industrial biomineralization of uranium is an important topic for SRB [66]. Washed cells of *D. desulfuricans* reduced soluble uranyl, UO_4^{2-}; U(VI), to insoluble uraninite, UO_2, U(IV), with lactate or H_2 as the electron donor [59]. The reduction of U (VI) to U(IV) has been demonstrated with *D. desulfuricans* in stationary phase following growth on lactate-sulfate medium [105] and in column studies using immobilized *D. desulfuricans* [96, 98]. Reduction of U(VI) by *D. desulfuricans* is shown in the electron micrographs in Fig. 4.10. Other reports of U(VI) reduction by SRB include the following: *Desulfomicrobium norvegicum* (formerly *D. baculatus*) [57]; *Desulfosporosinus orientis* and *Desulfosporosinus* sp. P3 [89]; *D. baarsii*, *D. sulfodismutans* and *D. vulgaris* [57]. *Desulfovibrio* sp. UFZ B 490 and *Desulfotomaculum reducens* are the only SRB that have been reported to grow with U(VI) as the electron acceptor [77, 78, 91].

Purified cytochrome c_3 from *D. vulgaris* was demonstrated to function as a uranium reductase [57]. Using mutants of *D. alaskensis* G20 lacking cytochrome c_3 and comparing the rate of U(VI) to wild type cells, the activity of U(VI) reduction was markedly inhibited with pyruvate or lactate as the electron donor and relatively

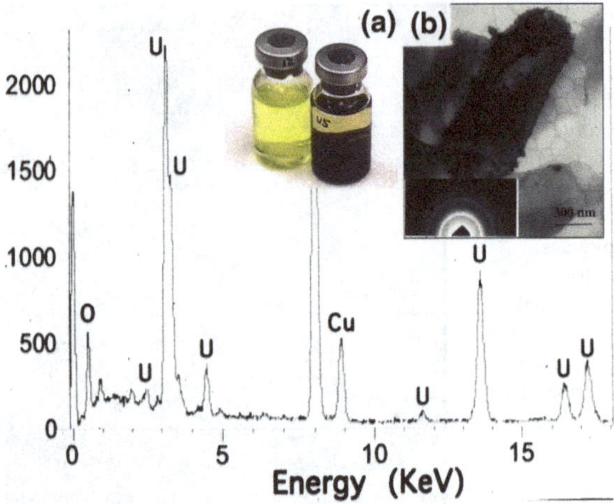

Fig. 4.10 EDX microanalysis of uranium precipitates associated with cells of *D. desulfuricans* grown for 2 days in a lactate/sulfate medium containing1 mM uranyl acetate. Uranium peaks are highlighted in yellow. *Inset* (*A*) shows a bottle of uninoculated culture medium with uranyl acetate and no deposits present in the medium. The bottle inoculated with the bacterium shows *black deposits* in the culture after 2 days. *Insert* (*B*) is an electron diffraction pattern from extracellular uranium aggregates. *Insert* (*C*) indicates TEM image of uraninite nanocrystals produced by *D. desulfuricans*. Methods for cultivation and electron microscopy are described by Tucker et al. [96]

little reduction occurred with H_2 [73]. Thioredoxin from *D. alaskensis* G20 will reduce U(VI) in the presence of thioredoxin reductase and NADPH [44]. Many of the SRB contain thioredoxin. The role of thioredoxin in reduction of U(VI) by other SRB remains to be established.

4.6 Biogeochemical Interactions

4.6.1 Biosulfide and Mineral Development

Research by Kucha et al. [46] reveals the possibility that sulfate-reducing bacteria were active in the Bleiberg lead-zinc deposit in Austria. The authors indicate the morphological similarity of filaments and spherules of sphalerite, zinc-bearing calcite and pyrite, as observed by electron microscopy, with mineral deposits recently reported in biofilms of sulfate-reducing bacteria. Additionally, the $\delta^{34}S$‰ of sulfur in metal sulfides of the Bleiberg lead-zinc deposit suggests bacteria were important in ore deposition.

Using sulfate-reducing bacteria added to artificial sea water and H_2 or lactate as the electron donor, Bass Becking and Moore [11] demonstrated the reaction with hydrogen sulfide with numerous metals, metal oxides or metallic salts with the production of minerals. Biogenic sulfide reacted with metallic iron to produce ferrous sulfide, metallic zinc or smithonite produced sphalerite, silver carbonate produced argentite, covellite was produced from malachite and from chrysocotta, digenite from cuprous oxide, and galena from lead carbonate. In natural biofilms containing sulfate-reducing bacteria, sphalerite is produced [47]. The potential for sulfate-reducing bacteria to contribute to mineral development should not be underestimated.

4.6.2 Iron and Reduced Selenium Compounds

With H_2 as the electron donor, a strain of *D. desulfuricans* reduces selenite to Se(0) and selenide (Se^{2-}), the reduced selenium atoms becomes incorporated into pyrite as iron-bearing selenides [103]. The mechanism for this formation of modified pyrite with iron selenides was suggested to follow previous reports that sulfides (and in this case selenide) absorbed onto cell surfaces readily react with metal ions [34, 45]. It has been reported that Se(0)could combine with amorphous structures of FeS and FeS_2 to produce analogous structures of FeSe, FeSSe, and $FeSe_2$ [63]. This process of bacterial reduction of selenite could promote a stable distribution of selenium in the geological environment.

4.6.3 Mobility of Arsenic

As reviewed by Lear et al. [48], there is considerable interest in the release of arsenic from aquatic sediments. With the reduction of sulfate by SRB, the level of hydrogen sulfide in the environment increases. With *Desulfosporosinus auripigmenti* (formerly *Desulfotomaculum auripigmentum*; [87] in a controlled laboratory setting, arsenic trisulfide (As_2S_3) is precipitated [69]. In a column study simulating the environment, sand coated with As(III)-bearing ferrihydrite [Fe $(OH)_3$] was exposed to *D. vulgaris*, biosulfide mobilized arsenic [18]. This process released As(III) into the column effluent and it was attributed to replacement of Fe $(OH)_3$ with FeS, mackinawite, and to a lesser amount Fe_3O_4, magnetite. The dynamic activity involving mineral presence provides for development of interesting models.

4.7 Summary and Perspective

This review has discussed the metabolism of metalloids, transition metals and actinides in the sulfate-reducing bacteria. These elements can be reduced, methylated, oxidized or precipitated by biotic and abiotic mechanisms. The inherent production of hydrogen sulfide can precipitate some of these metals as insoluble metal sulfides. Those elements that are redox active can be reduced by the sulfide, quite often forming insoluble salts. The oxidation of those found in the elemental form can be accelerated by both, the presence of sulfide and the consumption of hydrogen gas evolving from the metal. The redox chemistry of the elements can also be coupled to the flow of electrons from cellular metabolism. This process involves substrate- or non-substrate specific enzymatic enzymes that are respiratory components of the cells. While cytochrome c_3 has been demonstrated to reduce specific metallic elements, enzymes and electron carriers with metal cofactors may also have the capability to reduce metal(loid)s. The cytochromes from sulfate/sulfur reducers are unique and it has been suggested by Lojou et al. [55], Assfalg et al. [2] that cytochromes (i.e., c_3 and c_7), with the bishistidinyl heme iron coordination, act as metal reducers. The interaction of a detoxifying protein with many different metals has potential practical applications. Such a system would be available for a number of detoxification situations. The sulfate-reducing bacteria are ubiquitous and as such, they are found in the human large intestine [82]. The understanding of how bacteria regulate the transformation of metal(loid)s can lead to the prevention and control of acute and chronic exposure to metal(loid)s in the human diet.

Disclaimer The findings and conclusions in this report are those of the author(s) and do not necessarily represent the views of the Agency for Toxic Substances and Disease Registry.

References

1. Abdelous A, Gong WL, Lutze W, Shelnutt JA, Franco R, Moura I (2000) Using cytochrome c_3 to make selenium wires. Chem Mat 12:1510–1512
2. Assfalg M, Bertini I, Bruschi M, Michel C, Turano P (2002) The metal reductase activity of some multiheme cytochromes *c*: NMR structural characterization of the reduction of chromium (VI) to chromium (III) by cytochrome c_7. Proc Nat Acad Sci US 99:9750–9754
3. Azabou S, Mechichi T, Sayadi S (2007) Zinc precipitation by heavy-metal tolerant sulfate-reducing bacteria enriched on phosphogypsum as a sulfate source. Minerals Eng 20:173–178
4. Barton LL, Tomei FA (1994) Application of microbial remediation in detoxification of sites containing metal wastes. In: Bhada RK, Ghassemi A, Ward TJ, Jamshidi M, Shahinpoor M (eds) Waste management: from risk to remediation, vol 1. ECM Press, Albuquerque, pp 267–285
5. Barton LL, Choudhury K, Thomson BM, Steenhoudt K, Groffman AR (1996) Bacterial reduction of soluble uranium: The first step on in situ immobilization of uranium. Radioact Waste Manage Environ Restor 20:141–151
6. Barton LL, Fardeau M-L, Fauque GD (2014) Hydrogen sulfide. A toxic gas produced by dissimilatory sulfate and sulfur reduction and consumed by microbial oxidation. In: Sigel A, Sigel H, Sigel RKO (eds) Metal ions in life sciences. Springer Science & Business Media, B. V., Dordrecht, pp 237–277
7. Barton LL, Tomei-Torres FA, Xu H, Zocco T (2014) Nanoparticles formed in microbial metabolism of metals and minerals. In: Barton LL, Xu H, Bazylinski D (eds) Nanomicrobiology: Physiological and Environmental Characteristics. Springer, Dordrecht, pp 147–176

8. Barton LL, Plunkett RM, Thomson BM (2003) Reduction of metals and nonessential elements by anaerobes. In: Ljungdahl LG, Adams MW, Barton LL, Ferry JG, Johnson MK (eds) Biochemistry and physiology of anaerobic bacteria. Springer, New York, pp 220–234
9. Barton LL, Goulhen F, Bruschi M, Woodards NA, Plunkett RM, Rietmeijer FJM (2007) The bacterial metallome: composition and stability with specific reference to the anaerobic bacterium *Desulfovibrio desulfuricans*. Biometals 20:291–302
10. Barton LL, Fauque GD (2009) Biochemistry, physiology and biotechnology of sulfate-reducing bacteria. Adv Appl Microbiol 68:41–98. doi:10.1016/S0065-2164(09)01202-7
11. Bass Becking LGM, Moore D (1961) Biogenic sulfides. Econ Geol 56:259–272
12. Beech IB, Sunner JA (2007) Sulphate-reducing bacteria and their role in corrosion of ferrous materials. In: Barton LL, Hamilton WA (eds) Sulphate-reducing bacteria Cambridge University Press, Cambridge, UK, pp 459–482
13. Biswas KC, Barton LL, Tsui KL, Shuman K, Gillespie J, Eze CS (2011) A novel method for the measurement of elemental selenium produced by bacterial reduction of selenite. J Microbiol Methods 86:140–144
14. Biswas KC, Woodards NA, Xu H, Barton LL (2009) Reduction of molybdate by sulfate-reducing bacteria. Biometals 22:131–139
15. Blessing TC, Wielinga BW, Morra MJ, Fendorf S (2001) Co^{III} EDTA- Reduction by *Desulfovibria vulgaris* and propagation of reactions involving dissolved sulfide and polysulfides. Environ Sci Technol 35:1599–1603
16. Bruschi M, Barton LL, Goulhen F, Plunkett RM (2007) Enzymatic and genomic studies on the reduction of mercury and selected metallic oxyanions by sulphate-reducing bacteria. In: Barton LL, Hamilton WA (eds) Sulphate-reducing bacteria Cambridge University Press, Cambridge, UK, pp 435–457
17. Burdige DJ, Nealson KH (1985) Chemical and microbiological studies of sulfate-mediated manganese reduction. Geomicrobiol J 4:361–387
18. Burton ED, Johnson SG (2012) The impact of microbial sulfate reduction on subsurface arsenic mobility. In: Ng NC, Noller BN, Naidu L, Bundschuh J, Bhattacharya P (eds) Understanding the geological and medical interface of arsenic. CRC Press, Boca Raton
19. Cataldo DA, Garland TR, Wildung RE, Fellows RJ (1989) Comparative metabolic behaviour and interrelationships of Tc and S in soybean plants. Health Phys 57:281–288
20. Chardin B, Giudici-Orticoni M-T, De Luca G, Guigliarelli B, Bruschi M (2003) Hydrogenases in sulfate-reducing bacteria function as chromium reductase. Appl Microbiol Biotechnol 63:315–326
21. Chen G, Ford TE, Clayton CR (1998) Interaction of sulfate-reducing bacteria with molybdenum dissolved from sputter-deposited molybdenum thin films and pure molybdenum powder. J Colloid Interface Sci 204:237–246
22. Choi SC, Bartha R (1993) Cobalamin-mediated mercury methylation by *Desulfovibrio desulfuricans* LS. Appl Environ Microbiol 59:290–295
23. Creamer NJ, Baxter-Plant VS, Henderson J, Potter M, Macaskie LE (2006) Palladium and gold removal and recovery from precious metal solutions and electronic scrap leachates by *Desulfovibrio desulfuricans*. Biotechnol Lett 28:1475–1484
24. Csotonyl JT, Stackebrandt E, Yurkov V (2006) Anaerobic respiration on tellurate and other metalloids in bacteria from hydrothermal vent fields in the Eastern Pacific Ocean. Appl Environ Microbiol 72:4950–4956
25. De Luca G, De Philip P, Dermoun Z, Rousset M, Vermeglio A (2001) Reduction of Technetium (VII) by *Desulfovibrio fructosivorans* is mediated by the nickel-iron hydrogenase. Appl Environ Microbiol 67:4583–4587
26. Deplanche K, Macaskie LE (2008) Biorecovery of gold by *Escherichia coli* and *Desulfovibrio desulfuricans*. Biotechnol Bioeng 99:1055–1064
27. Dilworth GL, Bandurski RS (1977) Activation of selenate by adenosine 5′-triphosphate sulphurylase from *Saccharomyces cerevisiae*. Biochem J 163:521–529
28. Ehrlich HL, Newman DK (2009) Geomicrobiology, 5th edn. CRC Press, Boa Raton

29. Ekstrom E, Morel F, Benoit J (2003) Mercury methylation independent of the acetyl-coenzyme a pathway in sulfate-reducing bacteria. Appl Environ Microbiol 69:5414–5422
30. El Bayoumy MA, Bewtra JK, Ali HI, Biswas N (1996) Biosorption of lead by biomass of sulfate reducing bacteria. Can J Civil Engr 24:840–843
31. Enning D, Venzlaff H, Garrelfs J, Dinh HT, Meyer V, Mayrhofer K, Hassel AW, Stratmann M, Widdell F (2012) Marine sulfate-reducing bacteria cause serious corrosion on iron under electroconductive biogenic mineral crust. Environ Microbiol 14:1772–1787
32. Fauque G, LeGall J, Barton LL. 1991. Sulfate-reducing and sulfur-reducing bacteria. In: Shively JM Barton LL (eds) Variations in autotrophic life. Academic Press, San Diego, CA, pp 271–337
33. Fauque GD, Barton LL (2012) Hemoproteins in dissimilatory sulfate- and sulfur-reducing prokaryotes. Adv Microbiol Phys 60:1–91
34. Fortin D, Ferris FG, Beveridge TJ (1997) Surface-mediated mineral development by bacteria. Am Rev Mineral 35:161–180
35. Fude L, Harris B, Urrutia MM, Beveridge TJ (1994) Reduction of Cr(VI) by a consortium of sulfate-reducing bacteria (SRB III). Appl Environ Microbiol 60:1525–1531
36. Gadd GM (2010) Metals, minerals and microbes: geomicrobiology and bioremediation. Microbiology 156:609–643
37. Geiger SL, Ross TJ, Barton LL (1993) Environmental scanning electron microscope evaluation of crystal and plaque formation associated with biocorrosion. Microscopy Res Tech 25:429–443
38. Gilmour CC, Elias DA, Kucken AM, Brown SD, Palumbo AV, Schadt CW, Wall JD (2011) Sulfate-reducing bacterium *Desulfovibrio desulfuricans* ND132 as a model for understanding bacterial mercury methylation. Appl Environ Microbiol 77:3938–3951. doi:10.1128/AEM.02993-10
39. Graham AM, Bullock AL, Maizel AC, Elias DA, Gilmour CC (2012) Detailed assessment of the kinetics of Hg-cell association, Hg methylation, and methylmercury degradation in several *Desulfovibrio* species. Appl Environ Microbiol 78:7337–7346
40. Humphries AC, Macaskie LE (2005) Reduction of Cr(VI) by palladized biomass of *Desulfovibrio vulgaris* NCIMB 8303. J Chem Technol Biotechnol 80:1378–1382
41. Humphries AC, Nott KP, Hall LD, Macaskie LE (2005) Reduction of Cr(VI) by immobilized cells of *Desulfovibrio vulgaris* NCIMB 8303 and *Microbacterium* sp. NCIMB 13776. Biotechnol Bioeng 90:589–596
42. Hockin SL, Gadd GM (2007) Bioremedation of metals and metalloids by precipitation and cellular binding. In: Barton LL, Hamilton WA (eds) Sulphate-reducing bacteria. Cambridge University Press, Cambridge, pp 405–434
43. Li X, Krumholz LR (2007) Regulation of arsenate resistance in *Desulfovibrio desulfuricans* G20 by an *arsRBSS* operon and an *arsC* gene. J Bacteriol 189:3705–3711
44. Li X, Krumholz LR (2009) Thioredoxin is involved in U(VI) and Cr(VI) reduction in *Desulfovibrio desulfuricans* G20. J Bacteriol 191:4924–4933
45. Kohn MJ, Riciputi LR, Stakes D, Orange DL (1998) Sulfur isotope variability in biogenic pyrite: reflections of heterogeneous bacterial colonization? Am Mineral 83:1454–1468
46. Kucha HK, Schroll E, Stumpfl EF (2005) Fossil sulfate-reducing bacteria in the Bleiberg lead-zinc deposit, Austria. Miner Deposita 40:123–126
47. Labrenz M, Druschel GK, Thomsen-Ebert T, Gilbert B, Welch SA, Kemner KM, Logan GA, Summons RE, De Stasio G, Bond PL, Lai B, Kelly SD, Banfield JF (2000) Formation of sphalerite (ZnS) deposits in natural biofilms of sulfate-reducing bacteria. Science 290:1744–1747
48. Lear G, Song B, Gault AG, Polya DA, Lloyd RJ (2007) Molecular analysis of arenate-reducing bacteria within Cambodian sediments following amendment with arsenate. Appl Env Microbiol 73:1041–1048
49. Lengke M, Southam G (2006) Bioaccumulation of gold by sulfate-reducing bacteria cultured in the presence of gold(I)-thiosulfate complex. Geochem Cosmochim Acta 70:3646–3661

50. Lloyd JR (2003) Microbial reduction of metals and radionuclides. FEMS Microbiol Rev 27:411–425
51. Lloyd JR, Yong P, Macaskie LE (1998) Enzymatic recovery of elemental palladium by using sulfate-reducing bacteria. Appl Environ Microbiol 64:4608–4609
52. Lloyd JR, Mabbett AN, Williams DR, Macaskie LE (2001) Metal reduction by sulphate-reducing bacteria: physiological diversity and metal specificity. Hydrometallurgy 59:327–337
53. Lloyd JR, Ridley J, Khizniak T, Lyalikova NN, Macaskie LE (1999) Reduction of technetium by *Desulfovibrio desulfuricans*: biocatalyst characterization and use in a flowthrough bioreactor. Appl Environ Microbiol 65:2691–2696
54. Lloyd JR, Thomas GH, Finlay JA, Cole JA, Macaskie LE (1999) Microbial reduction of technetium by *Escherichia coli* and *Desulfovibrio desulfuricans*: enhancement via the use of high-affinity strains and effect of process parameters. Biotechnol Bioeng 66:122–130
55. Lojou E, Bianco P, Bruschi M (1998) Kinetic studies on the electron transfer between bacterial c-type cytochromes and metal oxides. J Electroanalytical Chem 452:167–177
56. Lovley DR (1993) Dissimilatory metal reduction. Ann Rev Microbiol 47:263–290
57. Lovley DR, Roden EE, Phillips EJP, Woodward JC (1993) Enzymatic iron and uranium reduction by sulfate-reducing bacteria. Marine Geol 113:41–53
58. Lovley DR (1995) Microbial reduction of iron, manganese, and other metals. Adv Agr 54:175–231
59. Lovley DR, Phillips EJP (1992) Reduction of uranium by *Desulfovibrio desulfuricans*. Appl Environ Microbiol 58:850–856
60. Lovley DR, Phillips EJP (1994) Reduction of chromate by *Desulfovibrio vulgaris* and its c_3 cytochrome. Appl Environ Microbiol 60:726–728
61. Lyalikova NN, Yurkova NA (1992) Role of microorganisms in vanadium concentrations and dispersion. Geomicrobiol J 10:15–26
62. Macy JM, Santini JM, Pauling BV, O'Neill AH, Sly LI (2000) Two new arsenate/sulfate-reducing bacteria: mechanisms of arsenate reduction. Arch Microbiol 173:49–57
63. Masschelein PH, Delaune RD, Patrick Jr WH (1990) Transformations of selenium as affected by sediment oxidation-reduction potential and pH. Environ Sci Technol 2491–96
64. Michel C, Giudici-Orticoni M-T, Bayman F, Bruschi M (2003) Bioremediation of chromate by sulfate-reducing bacteria, cytochromes c_3 and hydrogenases. Water Air Soil Pollut Focus 3:161–171
65. Mikheenko IP, Rousset M, Dementin S, Macaskie LE (2008) Bioaccumulation of palladium by *Desulfovibrio fructosivorians* wild-type and hydrogenase-deficient strains. Appl Environ Microbiol 74:6144–6146
66. Min MZ, Xu HF, Barton LL, Wang RC, Peng XJ, Wiatrowski W (2005) Biomineralization of uranium: a simulated experiment and its significance. Acta Geologica Sinica 79:134–138 (English edition)
67. Moll H, Merroun ML, Hennig Ch, Rossberg A, Selenska-Pobell S, Bernard G (2006) The interactions of *Desulfovibrio äspöensis* DSM 10613 with plutonium. Radiochem Acta 94:815–824
68. Neria-Gonzalez I, Wang ET, Ramirez F, Romero JM, Hernandez-Rodriguez C (2006) Characterization of bacterial community associated to biofilms of corroded oil pipelines from the southeast of Mexico. Anaerobe 12:122–133
69. Newman DK, Beveridge TJ, Morel FMM (1997) Precipitation of arsenic trisulfide by *Desulfotomaculum auripigmentum*. Appl Environ Microbiol 63:2022–2028
70. Odom JM (1993) Industrial and environmental activities of sulfate-reducing bacteria. In: Odom JM, Singleton R Jr (eds) The sulfate-reducing bacteria: contemporary perspectives. Springer, New York, pp 189–210
71. Painter EP (1941) The chemistry and toxicity of selenium compounds, with special reference to the selenium problem. Chem Rev 28:179–213

72. Parks JM, Johs A, Podar M, Bridou R, Hurt RA Jr, Smith SD, Tomanicek SJ, Qian Y, Brown SD, Bramdt CC, Palumbo AV, Smith JC, Wall JG, Elias DA, Liang L (2013) The genetic basis for bacterial mercury methylation. Science 339:1332–1335
73. Payne RB, Darren M, Gentry DM, Rapp-Giles BJ, Casalot L, Wall J (2002) Uranium reduction by *Desulfovibrio desulfuricans* strain 20 and a cytochrome c_3 mutant. Appl Environ Microbiol 68:3129–3132
74. Peck HD Jr, LeGall J (1994) Inorganic microbial sulfur metabolism. Meth Enzymol 243:1–682
75. Peretrukhin VF, Khizhnyak TV, Lyalikova NN, German KE (1996) Biosorption of technetium-99 and some actinides by bottom sediments of Lake Beloe Kosino of the Moscow region. Radiochemistry 38:440–443
76. Pérez RM, Cabrera G, Gómez JM, Abalos A, Cantero D (2010) Combined strategy for the precipitation of heavy metals and biodegradation of petroleum in industrial wastewaters. J Hazardous Material 182:96–902
77. Pietzsch K, Babel W (2003) A sulfate-reducing bacterium that can detoxify U(VI) and obtain energy by nitrate reduction. J Basic Microbiol 43:348–361
78. Pietzsch K, Hard BC, Babel W (1999) A *Desulfovibrio* capable of growing by reducing U (VI). J Basic Microbiol 39:365–372
79. Postgate JR (1979) The sulphate-reducing bacteria. Cambridge University Press, Cambridge
80. Poulain AJ, Barkay T (2013) Cracking the mercury methylation code. Science 339:1280–1281
81. Rashamuse KJ, Whiteley CG (2007) Bioreduction of Pt(IV) from aqueous solution using sulphate-reducing bacteria. Appl Microbiol Biotechnol 75:1429–1435
82. Rey FE, Gonzales MD, Cheng J, Wu M, Ahern PP, Gordon JI (2013) Metabolic niche of a prominent sulfate-reducing human gut bacterium. PNAS 110:13582–13587
83. Sass H, Cypionka H (2004) Isolation of sulfate-reducing bacteria from the terrestrial deep subsurface and description of *Desulfovibrio cavernae* sp. nov. Sys Appl Microbiol 27:541–548
84. Sass H, Ramamoorthy S, Yarwood C, Langner H, Schumann P, Kroppenstedt RM, Spring S, Rosenberg RF (2009) *Desulfovibrio idahonensis* sp. nov., sulfate-reducing bacteria isolated from a metal(loid)-contaminated fresh water sediment. Internat J System Evol Microbiol 59:2208–2214
85. Smith W-L (2001) Hexavalent chromium reduction and precipitation by sulphate-reducing bacterial biofilms. Environ Geochem Health 23:297–300
86. Spallholz JE (1994) On the nature of selenium toxicity and carcinostatic activity. Free Radic Biol Med 17:45–64
87. Stackebrandt E, Schumann P, Schüler E, Hippe H (2003) Reclassification of *Desulfotomaculum auripigmentum* as *Desulfosporosinus auripigmenti* corrig., comb. nov. Int J Syst Evol Microbiol 53:1439–1443
88. Stolz JF, Oremland RS (1999) Bacterial respiration of arsenic and selenium. FEMS Microbiol Rev 23:615–627
89. Suzuki Y, Kelly SD, Kemner KM, Banfield JF (2004) Enzymatic U(VI) reduction by *Desulfosporosinus* species. Radiochim Acta 92:11–16
90. Taylor DE (1999) Bacterial tellurite resistance. Trends Microbiol 7:111–115
91. Tebo BM, Obraztsova AY (1998) Sulfate-reducing bacterium with Cr(VI), U(VI), Mn(IV), Fe(III) as electron acceptors. FEMS Microbiol Lett 162:195–198
92. Teekayuttasakul P, Annachhatre AP (2008) Lead removal and toxicity reduction from industrial wastewater through biological sulfate reduction process. J Environ Sci Health A Tox Haz Subst Environ Eng 43:1424–1430
93. Teixeira M, Moura I, Xavier AV, DerVartanian DV, LeGall J, Peck HD Jr, Huynh BH, Moura JJG (1983) *Desulfovibrio gigas* hydrogenase: Redox properties of the nickel and iron-sulfur centers. Eur J Biochem 130:481–484

94. Tomei FA, Barton LL, Lemanski CL, Zocco TG, Fink NH, Sillerud LO (1995) Transformation of selenate and selenite to elemental selenium by *Desulfovibrio desulfuricans*. J Ind Microbiol 14:329–336
95. Tomei FA, Mitchell R (1986) Development of an alternative method for studying the role of H_2-consuming bacteria in the anaerobic oxidation of iron. In: Dexter SC (ed) Biologically induced corrosion. National Association of Corrosion Engineers, Houston, pp 309–320
96. Tucker MD, Barton LL, Thomson BM (1998) Reduction of Cr, Mo, Se and U by *Desulfovibrio desulfuricans* immobilized in polyacrylamide gels. J Ind Microbiol Technol 20:13–19
97. Tucker MD, Barton LL, Thomson BM (1997) Reduction and immobilization of molybdenum by *Desulfovibrio desulfuricans*. J Environ Qual 26:1146–1152
98. Tucker MD, Barton LL, Thomson BM (1996) Kinetic coefficients for simultaneous reduction of sulfate and uranium by *Desulfovibrio desulfuricans*. Appl Microbiol Biotechnol 46:74–77
99. Tuttle JH, Dugan PR, Randles CI (1969) Microbial sulfate reduction and its potential utility as an acid mine water pollution abatement procedure. Appl Microbiol 17:297–302
100. Vandieken V, Knoblauch C, Jørgensen BB (2006) *Desulfovibrio frigidus* sp. nov. and *Desulfovibrio ferrireducens* sp. nov. psychrotolerant bacteria isolated from Artic fjord sediments (Svalbard) with the ability to reduce Fe(III). Int J Syst Evol Microbiol 56:681–685
101. Wall JD, Krumholz LR (2006) Uranium reduction. Ann Rev Microbiol 60:149–166
102. Woolfolk CA, Whiteley HR (1962) Reduction of inorganic compounds with molecular hydrogen by *Micrococcus lactilyicus*. J Bacteriol 84:648–658
103. Xia X, Enokida Y, Sawada K, Ohnuki T (2007) Bioreduction of selenium by sulfate reducing bacterium and its influence on selenium transport in geological formations. Proc Int Symp EcoTropia Science. ISETS07(2007)
104. Xu H, Barton LL (2013) Se-bearing colloidal particles produced by sulfate-reducing bacteria and sulfide oxidizing bacteria: a TEM study. Adv Microbiol 3:205–211. doi:10.4236/aim.2013.32031
105. Xu H, Barton LL, Zhang P, Wang Y (1999) TEM investigation of U6+ and Re7+ reduction by *Desulfovibrio desulfuricans*, a sulfate-reducing bacterium. Sci Basis Nuclear Waste Manage 23:299–304
106. Yong P, Farr JPG, Harris IR, Macaskie LE (2002) Palladium recovery by immobilized cells of *Desulfovibrio desulfuricans* using hydrogen as the electron donor in a novel electrobioreactor. Biotechnol Lett 24:205–212

Index

D. Saffarini (ed.), *Bacteria-Metal Interactions*,
DOI 10.1007/978-3-319-18570-5

Zeitfracht Medien GmbH
Ferdinand-Jühlke-Straße 7
99095 Erfurt, Deutschland
produktsicherheit@kolibri360.de